D1754602

Existing Sewer Evaluation and Rehabilitation

ASCE Manuals and Reports on Engineering Practice No. 62
WPCF Manual of Practice No. FD-6

Prepared by a Joint Task Force of the American
Society of Civil Engineers and the Water
Pollution Control Federation

The Joint Task Force was composed of members of the ASCE Environmental Engineering Division Committee on Water Pollution Management and the WPCF Facilities Development Subcommittee of the Technical Practice Committee.

Published by the

American Society of Civil Engineers
345 East 47th Street
New York, New York 10017

and the

Water Pollution Control Federation
2626 Pennsylvania Avenue, N.W.
Washington, D.C. 20037

Manuals of Practice
for
Water Pollution Control

The WPCF Technical Practice Committee (formerly the Committee on Sewage and Industrial Wastes Practice of the Federation of Sewage and Industrial Wastes Associations) was created by the Federation Board of Control on October 11, 1941. The primary function of the committee is to originate and produce, through appropriate subcommittees, special publications dealing with technical aspects of the broad interests of the Federation. These manuals are intended to provide background information through a review of technical practices and detailed procedures that research and experience have shown to be functional and practical.

The material presented in this publication has been prepared in accordance with generally recognized engineering principles and practices, and is for general information only. This information should not be used without first securing competent advice with respect to its suitability for any general or specific application.

The contents of this publication are not intended to be and should not be construed to be a standard of the American Society of Civil Engineers (ASCE) or the Water Pollution Control Federation (WPCF) and are not intended for use as a reference in purchase specifications, contracts, regulations, statutes, or any other legal document.

No reference made in this publication to any specific method, product, process, or service constitutes or implies an endorsement, recommendation, or warranty thereof by ASCE or WPCF.

ASCE and WPCF make no representation or warranty of any kind, whether express or implied, concerning the accuracy, completeness, suitability or utility of any information, apparatus, product, or process discussed in this publication, and assume no liability therefor.

Anyone utilizing this information assumes all liability arising from such use, including but not limited to infringement of any patent or patents.

Water Pollution Control Federation
Technical Practice Committee
Control Group

R. R. Rimkus, *Chairman*
S. P. Graef, *Vice-Chairman*

H. Miller
M. C. Mulbarger

T. M. Regan
C. S. Zickefoose

Authorized for Publication by the Board of Control
Water Pollution Control Federation

1983

Robert A. Canham, *Executive Director*

Joint Task Force on Existing Sewer Evaluation and Rehabilitation

Dan L. Glasgow, Chairman*
Paul M. Beall
Phillip M. Botch
Herman Brinkmann
Carl A. Brunner*
Samar Chatterjee
Lloyd K. Clark
Jerry R. Dailey
Richard Field
James H. Forbes
Alberto F. Gutierrez*
Akhtar Hamid
Dwight Hensley
Kenneth T. Holmes*

J. Thomas Jacobs
Douglas Jacobson
Larry P. Jaworski
Robert C. Mayeron
Lyndel W. Melton
Richard L. Nogaj*
Warren W. Sadler*
B. J. Schrock
Stephen R. Maney
R. John Tagg
Edward L. Tharp*
William A. Thoden
Richard O. Thomasson
Robert Turkeltaub

Stephen Waring

*Principal Contributing Author

In addition to the Task Force, an advisory subcommittee for product manufacturers which reviewed this manual under the direction of the Task Force Chairman included:

Robert J. Barletta
Mike Bealey
William J. Clarke
A. B. Colthorp
Kenneth Kienow

Barry W. Kostyk
Craig Marley
Stan Mruk
William D. Nesbeitt
Gerald E. Scheid

This manual was also reviewed by an editorial subcommittee which included:

Dan L. Glasgow, Chairman
Malise J. Graham

Gordon S. Magnuson
Charles A. Parthum

B. J. Schrock

Thanks are due to the WPCF Metrication Committee, M. J. Graham, Chairman, and to Richard H. Sullivan, American Public Works Association, for the contributions to this manual.

Staff assistance was provided by Eugene De Michele and Berinda J. Ross for the Water Pollution Control Federation and Harry N. Tuvel for the American Society of Civil Engineers.

MANUALS AND REPORTS ON ENGINEERING PRACTICE

(As developed by the ASCE Technical Procedures Committee,
July 1930, and revised March 1935, February 1962, April 1982)

A manual or report in this series consists of an orderly presentation of facts on a particular subject, supplemented by an analysis of limitations and applications of these facts. It contains information useful to the average engineer in his everyday work, rather than the findings that may be useful only occasionally or rarely. It is not in any sense a "standard," however; nor is it so elementary or so conclusive as to provide a "rule of thumb" for nonengineers.

Furthermore, material in this series, in distinction from a paper (which expresses only one person's observations or opinions), is the work of a committee or group selected to assemble and express information on a specific topic. As often as practicable the committee is under the general direction of one or more of the Technical Divisions and Councils, and the product evolved has been subjected to review by the Executive Committee of that Division or Council. As a step in the process of this review, proposed manuscripts are often brought before the members of the Technical Divisions and Councils for comment, which may serve as the basis for improvement. When published, each work shows the names of the committees by which it was compiled and indicates clearly the several processes through which it has passed in review, in order that its merit may be definitely understood.

In February 1962 (and revised in April, 1982) the Board of Direction voted to establish:

A series entitled 'Manuals and Reports on Engineering Practice,' to include the Manuals published and authorized to date, future Manuals of Professional Practice, and Reports on Engineering Practice. All such Manual or Report material of the Society would have been refereed in a manner approved by the Board Committee on Publications and would be bound, with applicable discussion, in books similar to past Manuals. Numbering would be consecutive and would be a continuation of present Manual numbers. In some cases of reports of joint committees, bypassing of Journal publications may be authorized.

Abstract

Existing Sewer Evaluation and Rehabilitation

This manual provides guidelines for the evaluation and rehabilitation of sanitary sewers.

The initial chapter introduces the purpose and scope of sanitary sewer rehabilitation. Subsequent chapters provide sufficient detail and information in such a manner that users of the manual may understand, evaluate and institute a rehabilitation program. Charts, figures and illustrations are used where practical to reinforce the text.

The major emphasis of the manual is on infiltration and inflow reduction with lesser emphasis being placed on maintaining the structural integrity of the sanitary sewer. Many of the rehabilitation methods presented are applicable for both aspects of sanitary sewer rehabilitation. The structural integrity aspects of sanitary sewer rehabilitation cannot be overemphasized because of the potentially destructive and costly events which often result after the collapse within the sewer system.

Copyright © 1983 by the Water Pollution Control Federation
Washington, D.C. 20037 U.S.A.

Library of Congress Catalog No. 65-5192

ISBN 0-943244-43-9

Printed in U.S.A. 1983
by
Lancaster Press, Inc.
Lancaster, Pa.

Composed by
Techna Type, Inc.
York, Pa.

Contents

Chapter 1	**INTRODUCTION** Overview and Historical Background Purpose and Scope The Need for Guidelines Format of the Manual	1
Chapter 2	**METHODS OF SEWER EVALUATION** Purpose of Sewer System Evaluation System Problems Evaluation Techniques Evaluation of Infiltration Evaluation of Inflow and Rainfall-Induced I/I Evaluation of Physical Conditions References	5
Chapter 3	**FLOW MONITORING** General Data Needs Flow Measurement Techniques Design of a System Monitoring Program References	43
Chapter 4	**METHODS OF SEWER REHABILITATION** General Considerations Pipeline Rehabilitation Manhole Rehabilitation Service Connection Rehabilitation References	63
Chapter 5	**MATERIALS USED FOR SEWER REHABILITATION** Manhole Rehabilitation Materials Materials for Rehabilitation of Main Sewer Line References	83
Chapter 6	**EFFECTIVENESS OF SEWER REHABILITATION** Measurement of Effectiveness of I/I Control Studies of Rehabilitation Effectiveness for I/I Control Expected Effectiveness of Sewer Rehabilitation Continuing Sewer Maintenance References	93

List of Tables

		Page
TABLE 2.1.	Inflow through manhole frames and covers.	27
TABLE 2.2.	Cost of rehabilitation for elimination of inflow.	37
TABLE 2.3.	Information gathered when examining for structural defects.	39
TABLE 4.1.	Estimated percentage of total infiltration attributed to building sewers.	80
TABLE 4.2.	Miscellaneous private property rehabilitation measures.	81

List of Figures

		Page
FIGURE 2.1	Typical weekly hydrograph for a monitoring site, South St. Paul (Dry Weather Flow, 5/29/81–6/5/81; Wet Weather Flow, 5/22/81–5/29/81).	10
FIGURE 2.2	Static groundwater gauge installation elevation.	11
FIGURE 2.3	Groundwater gauge installation detail.	12
FIGURE 2.4	Sample smoke testing notice.	13
FIGURE 2.5	Daily component flow comparison.	30
FIGURE 2.6	Hydrographs—the most common method of presenting flow data.	34
FIGURE 2.7	Cost curve for determination of I/I sources to be rehabilitated.	38
FIGURE 3.1	Flow monitoring weirs.	46
FIGURE 3.2	Weir details and installation.	47
FIGURE 3.3	Parshall flume.	48
FIGURE 3.4	Palmer-Bowlus flume details.	49
FIGURE 3.5	Palmer-Bowlus flume installation.	50
FIGURE 4.1	Section removed from a sewer pipeline showing a severe roof intrusion problem.	65
FIGURE 4.2	Chemical grout sealing.	67
FIGURE 4.3	Typical arrangement for applying chemical grout to small diameter pipe.	69
FIGURE 4.4	Typical arrangement for sealing large diameter pipe with grouting rings.	70
FIGURE 4.5	Sliplining installation methods.	73
FIGURE 4.6	Full encirclement tapping saddle.	74
FIGURE 4.7	Inversion lining installation procedure.	75
FIGURE 4.8	Manhole lid insert. The lip rests directly on the manhole frame. The manhole cover normally rests on top of the lip.	76
FIGURE 5.1	Manhole covers with concealed pickholes.	84
FIGURE 5.2	Manhole covers with self-sealing tapered groove and gasket.	84
FIGURE 5.3	Double ring placement of flexible rubber-like gasket material for adjustment of ring joints.	85
FIGURE 5.4	Double ring placement of flexible rubber-like gasket material for outside drop manhole joints.	85
FIGURE 5.5	Inclined adjustment rings.	85
FIGURE 5.6	Flexible manhole sleeve and flange.	86
FIGURE 5.7	Flexible manhole sleeves with flange cast in manhole wall.	86

Foreword

This is the first edition of the Manual of Practice on "Existing Sewer Evaluation and Rehabilitation" by the Joint ASCE-WPCF Task Force.

The Joint ACSE-WPCF Task Force was organized in 1981 by WPCF's Technical Practice Committee for the purpose of addressing the current state of technology of the evaluation and rehabilitation of existing sewers. The Task Force was composed of ASCE and WPCF members. Mr. Dan Glasgow served as the Chairman with Mr. B. J. Schrock as the Vice-Chairman. The Task Force was divided into six groups chaired by Mr. Warren W. Sadler, Mr. Alberto F. Gutierrez, Mr. Edward L. Tharp, Mr. Kenneth T. Holmes, Mr. Richard J. Nogaj, and Mr. Carl A. Brunner.

To broaden the base of experience to be reflected in the manual, each chapter was prepared by one of the six groups, comprised of Task Force members whose experience and background corresponds to the respective chapter. The manual has been reviewed by ASCE's Environmental Engineering Division and by the WPCF Technical Practice Committee Control Group. Both express their appreciation to the members of the Task Force for their contributions to the success of this project.

The purpose of this manual is to provide general guidance and to serve as a source of information in the evaluation and rehabilitation of existing sewers. The book provides information on conducting I/I analyses, sewer system evaluation surveys, and rehabilitation methodology. It contains neither standards nor rules and regulations. Rather, the design information presented is intended as technical guidance reflecting sound, professional practice.

Chapter 1
INTRODUCTION

1 Overview and Historical Background	2 The Need for Guidelines
1 Purpose and Scope	2 Format of the Manual

OVERVIEW AND HISTORICAL BACKGROUND

There are a number of purposes for sewer rehabilitation. The primary one in the past probably was to maintain the structural integrity of the sewer system for dependable transfer of wastewater from the source to the treatment plant or receiving water. If this were the only purpose, concern over effectiveness of wastewater transfer—other than for structural reasons—would be minimal. With the passage of Public Law 92-500 in 1972, however, far more emphasis was placed on sewer rehabilitation to reduce the hydraulic loads placed on treatment plants from excessive infiltration and inflow (I/I). The reduction of I/I can result in a significant reduction in hydraulic loading into collection and treatment facilities during periods of wet weather, thus lowering capital costs associated with oversized facilities, reducing operation and maintenance costs, and prolonging the life-capacity of the facility.

The U.S. Environmental Protection Agency's (EPA) requirements for I/I analysis when preparing facility plans are an indication of the degree of emphasis being placed on sewer rehabilitation. This manual is a guide to the individual who is confronted with evaluating and rehabilitating existing sewer facilities.

Recent technological advances, such as closed-circuit television cameras, have made it possible to observe the condition of existing sewers. This technology, in addition to improved analytical methods and materials, often has made rehabilitation of existing sewers a cost-effective alternative to relief sewer construction or replacement.

PURPOSE AND SCOPE

The purpose of this manual is to provide general guidance and to serve as a source of information in the evaluation and rehabilitation of existing sewers. Accordingly, typical procedures, case studies, and examples are presented to aid the user.

This book provides information on conducting I/I analyses, sewer system evaluation surveys, and rehabilitation methodology. It contains neither standards nor rules and regulations. Rather, the design information presented is intended as technical guidance reflecting sound, professional practice. The technologies discussed here were selected because of past experience and because of the availability of information and performance data. The exclusion of a particular method in this manual does not reflect on its acceptability. All available techniques should be considered when evaluating and planning for the reconditioning of existing sewers.

This manual discusses:

- Existing sewer system problems;
- Techniques for the evaluation of performance of existing sewers;
- Methods for evaluation of information gathered in testing;
- Techniques for measuring flow;
- The design of a system monitoring program;

EVALUATION/REHABILITATION

- A procedure for selecting the method for and overall approach to sewer rehabilitation;
- Advantages, limitations, disadvantages, costs, and feasibility of pipeline rehabilitation;
- Various methods of pipeline reconditioning;
- Rehabilitation for manholes;
- Materials used for rehabilitation of pipelines and manholes;
- The effectiveness of rehabilitation.

THE NEED FOR GUIDELINES

The regulatory guidelines to control sewer system/effluent, promulgated by PL 92-500, regarding study, operation, maintenance, and repair of sewer systems indicated that the technical guidelines available were for the most part inadequate.

The EPA evaluation of the success of sewer system rehabilitation with respect to I/I indicated that correlation of the diverse methods of pre-rehabilitation study and post rehabilitation evaluation (method) was virtually impossible. Thus, the need for state-of-the-art technical guidelines for use by professionals in the development of comprehensive sewer rehabilitation programs became evident.

Engineering practice in response to the regulatory guidelines has matured over the past 10 years. This state-of-the-art manual will maximize potential for sewer rehabilitation success, and provides a measure against which future rehabilitation methods may be compared.

FORMAT OF THE MANUAL

The information in this manual is intended for use by engineers, municipalities, regulatory agencies, and all others who have some responsibility for the planning, designing, constructing, financing, regulating, operating, or maintaining of sewer systems. Every effort has been made to address chronologically the development of analyzing the I/I, evaluating the sewer system surveys, and outlining the methods available for rehabilitation and their effectiveness.

The manual comments on three major topics: sewer evaluation, sewer rehabilitation, and the effectiveness of reconditioning sewers. Chapter 2 discusses techniques used for evaluation: the evaluation of the infiltration data, the location of the infiltration, the evaluation of rainfall-induced infiltration, the evaluation of inflow, and the evaluation of physical conditions.

Chapters 3 and 4 include descriptions of the various rehabilitation techniques that are currently utilized for sewer systems and considerations for the selection of a particular method, information on the rehabilitation of manholes, and a discussion of the problems and rehabilitation methods for service laterals.

Chapter 5 discusses materials used for manhole rehabilitation and the reconditioning of main sewer lines. The effectiveness of pipeline rehabilitation is discussed in Chapter 6.

Definitions of some of the terms used frequently throughout the manual may be helpful to the reader. The following definitions were extracted from the February 11, 1974, *Federal Register*, "Title 40 Rules and Regulations," Section 35.905.

Infiltration. The water entering a sewer system and service connections from the ground, through such means as, but not limited to, defective pipes, pipe joints, connections, or manhole walls. Infiltration does not include, and is distinguished from, inflow.

Inflow. The water discharged into a sewer system and service connections from such sources as, but not limited to, roof leaders, cellar, yard and area drains, foundation drains, cooling water discharges, drains from springs and swampy areas, manhole covers, cross connections from storm sewers and combined sewers, catch basins,

INTRODUCTION

storm water, surface runoff, street washes, or drainage. Inflow does not include, and is distinguished from, infiltration.

Infiltration/inflow. The total quantity of water from both infiltration and inflow without distinguishing the source.

Excessive infiltration/inflow. The quantities of infiltration/inflow which can be economically eliminated from a sewer system by rehabilitation, as determined by a cost-effectiveness analysis that compares the costs for transportation and treatment of the infiltration/inflow, subject to the provisions in Section 35.927 of the Federal Register.

Chapter 2
METHODS OF SEWER EVALUATION

5	Purpose of Sewer System Evaluation	32	Evaluation of Inflow and Rainfall-induced I/I
6	System Problems	38	Evaluation of Physical Conditions
7	Evaluation Techniques	41	References
27	Evaluation of Infiltration		

A serious problem results from excessive infiltration into sewers from groundwater sources, and high inflow rates into sewer systems directly from sources other than those that sewer conduits are intended to serve. The hydraulic and sanitary effects of these extraneous flows are particularly important because urban growth creates the need for all available sewer system capacity. The pollutional effects of bypassed, spilled, and undertreated wastewater flows caused by I/I are deterrents to the overall objective of protecting the nation's water resources.

PURPOSE OF SEWER SYSTEM EVALUATION

In the past decade, a determined effort has been made in the U.S. to reduce the effects of I/I on sewer systems. To assist engineers, municipalities, and regulatory agencies, EPA prepared guidelines for conducting I/I evaluations. That evaluation is divided into two parts: 1) the I/I analysis and 2) the sewer system evaluation survey (SSES).

The I/I analysis determines the existence or nonexistence of excessive infiltration and inflow. Correction of infiltration in existing sewer systems involves: 1) evaluation and interpretation of wastewater flow conditions to determine the presence and extent of excessive extraneous water flows from sewer system sources, 2) the location and gauging of such infiltration flows, 3) the elimination of these flows by various repair and replacement methods, and 4) a diligent, continuous maintenance program.

In the case of inflow conditions, the problem is two-faceted: prevention and cure. Prevention of excessive inflow volumes is a matter of regulating sewer uses and enforcement of applicable precepts and codes by means of information obtained through vigilant surveys and surveillance methods. Correction of existing inflow conditions involves discovery of points of inflow connections; determination of their legitimacy; assignment of the responsibility for correction of such conditions; establishment of inflow control policies where none have been in effect; and institution of corrective policies and measures, backed by investigative and enforcement procedures.

Control of I/I in all future sewer construction work, and the search for and correction of excessive intrusion of extraneous waters into existing sewer systems, is an essential part of sewer system management. A sewer system cannot be rehabilitated once and then be expected never

EVALUATION/REHABILITATION

to develop additional points of infiltration or inflow. A regular preventive maintenance program must be instituted to control extraneous water flows. Efforts to achieve higher standards of effluent quality through advanced degrees of treatment and funds dedicated to maintaining more rigid quality standards in public waters will be thwarted or rendered financially unsound if I/I is permitted to rob sewers of carrying capacities and treatment plants of their process performance capabilities.

To determine whether the I/I is excessive, information is gathered for making rough cost estimates for transportation and treatment of the I/I versus elimination through corrective action. If this initial analysis indicates that the I/I is excessive, the next phase should be the sewer system evaluation survey. The survey is the systematic examination of the sewer system to determine the specific location flow rate and rehabilitation costs of each I/I source. The sewer system evaluation survey is intended to confirm the general overall findings of the analysis program and to convey preliminary diagnoses into firm conclusions regarding the presence of, location, and degree of I/I. It must also determine what I/I intrusion is excessive or non-excessive, according to criteria stated in PL 92-500 and EPA rules and regulations. Definitive cost effectiveness studies supported by the actual findings of the evaluations survey are used to estimate the amount of I/I that could be eliminated as compared to the cost of the expanded physical facilities.

This conversion of preliminary findings into positive evaluation facts must be based on a detailed diagnosis of sewer system conditions. Such detail augments and supplements the more general data obtained during the analysis phase. Thus, the findings of the sewer system evaluation survey must dictate the nature of corrective actions, their costs, the means by which I/I will be controlled, and the basis for treatment plant capacity design decisions. The evaluation survey phase will determine the extent of sewer rehabilitation, in a rational sequence. An accomplishment factor must be considered because the estimates on flow that would be cost effective to keep may not be 100% accurate.

If the findings of the analysis stage clearly demonstrate that excessive I/I does not exist in the system, and if state and federal agencies concur with that conclusion, the evaluation phase should not be undertaken.

The evaluation survey must be planned and carried out to produce the type of authentic information that will justify the subsequent conclusions and recommendations. The depth and dimension of system evaluation must meet this criterion if it is to be the instrument for determining the actual sources of I/I, the scope of the problem, the means for correcting or alleviating it, the costs involved, and the determinant of the most cost-effective means for handling the problem in each specific system.

SYSTEM PROBLEMS

Infiltration and inflow seriously affect the operation of sewer systems and pumping, treatment, and overflow regulator facilities. It also adversely affects the urban environment and the quality of water resources. Some examples of the detrimental effects are: usurpation of sewer facility capacity that should be reserved for present sanitary wastewater flows and future urban growth; need for construction of relief sewer facilities before originally scheduled dates; surcharging and backflooding of sewers into streets and private properties; bypassing of raw wastewater at various points of spill or diversion into storm drains or nearby watercourses; surcharging of pump stations, resulting in excessive wear on equipment, higher power costs, or bypassing of flows to adjacent water sources; surcharging of wastewater treatment plants, with adverse consequences to treatment efficiency; diversion of flow from secondary-tertiary treatment stages, or bypassing of volumes of untreated wastewater into receiving waters; and increases in the incidence and duration of stormwater overflows at combined sewer regulators.

As defined in Chapter 1, the term *infiltration* covers the volume of groundwater entering sewer systems from the soil through defective joints, broken or cracked pipes, improperly made connections, manhole walls, or other means. Inflow differs in that it is the result of deliberately planned or expediently devised connections of sources of extraneous wastewater into sewer systems. These connections dispose of unwanted stormwater or other drainage water and wastes into a convenient drain conduit. This includes the deliberate or accidental draining of low-lying or flooded areas into sewer systems through manhole covers.

Rainfall-induced infiltration, a category between infiltration and inflow, is an increase in infiltration caused by rainfall-induced changes in subsurface conditions. Regardless of the source of waters that enter sewers and affect their ability to provide urban sanitation and drainage, the net result is the same: usurpation or reduction of valuable conduit and treatment capacities. The sewer systems so affected cannot distinguish between groundwaters that have infiltrated through defective points of entry, and those that have flowed into sewers from direct pipe connections.

EVALUATION TECHNIQUES

Flow components. Flow measurement, as a part of sewer system evaluation, is usually undertaken to define some variation of a certain flow component with time or to define peak and/or minimum flow conditions. Because flow measurement results may be used in design of capital improvements, it is critical to employ equipment and methods that minimize inaccuracies during flow measurements. It is also important to recognize the limitations of the data being collected. Flow measurement techniques are covered in Chapter 3.

Base flow, infiltration, and inflow determination. Flow in sanitary sewer systems consists of three components: base flow, infiltration, and inflow. Separation and quantification of these components is often a prime objective of flow metering. Base flow can be determined in several ways with varying degrees of accuracy. Water consumption data adjusted for seasonal peaks, irrigation, unmetered connections, and water meter inaccuracies, are often used. Another method of base flow determination entails the measurement of minimum flow rates to determine infiltration rates and then subtraction of this rate from metered flow during dry weather conditions. Per capita water consumption estimates for residences and commercial establishments upstream of the metering location can also be evaluated to determine base flow. Infiltration can be determined by subtracting base flow from total metered flow during dry weather or by compiling flow isolation measurements. Inflow is measured during wet weather conditions and is determined by subtracting base flow and infiltration from data recorded during wet weather conditions.

Peak rate determinations. Flow data for peak conditions are usually desired for each flow component. Base flow peaks can be obtained from recorded data by subtracting infiltration determined during dry weather conditions. Rates of infiltration should be determined if possible through correlation of groundwater flow data and past periods of high groundwater conditions.

Two peak rates of inflow usually should be established. The peak hour inflow rate is often required to size pump stations, interceptors, and other equipment that must handle wet weather surges. The peak daily flow rate may be used to size equalization basins and other flow storage or settling devices. Attempts have been made to correlate peak inflow rates with rainfall intensity.[1] If peak inflow rates are desired from gauging data, care must be taken to ensure that the monitoring site does not surcharge or, if this occurs, proper equipment is in use. All overflows or bypasses from the drainage area also must be monitored.

Annual volume determination. The determination of the annual volumes of base flow, infiltration, and inflow may be desired and can be predicted by evaluation of historical data in conjunction with results from a flow monitoring program. Base flow volume is often taken as some percentage of total potable water sales to allow for meter loss, sprinkling, and so forth. Care must be taken to include institutions or residences that use private sources of water but discharge to the sewer system. Infiltration volumes must be adjusted for seasonal fluctuations of groundwater levels. This can often be accomplished by comparing monthly water sales to wastewater flow records and developing a weighted distribution to reflect typical seasonal infiltration rates. Inflow volume can be projected proportionately using metered data and a known precipitation volume and obtaining the annual total precipitation.

Procedures. To determine the wastewater flow parameters discussed above, it is necessary to obtain wastewater flow data, rainfall data, groundwater data, and water consumption information. To determine infiltration and inflow for an existing sewer subsystem, flow-metering equipment must be in place, permanently or temporarily. In most sewer system evaluation studies, temporary equipment provides most flow information required to evaluate wastewater flow variations within sewer subsystems. Permanent metering installations at treatment facilities and pumping stations may provide additional flow data.

Pre-installation Considerations. Proper planning for flow metering is a critical step for any sewer evaluation monitoring program. The first consideration is the use of the data. Which parameters must be defined? How accurately must they be defined? What is the long-range impact on, for example, plant design or interceptor sizing?

The length of the metering period may have some bearing on equipment selection. Long-term metering may warrant the purchase of a portable flume or construction of a weir, whereas short-term metering may not. An item often overlooked is the method of data storage by the recorder. Several methods are available: circular and strip charts, magnetic tape, memory chips, and telemetering to a remote location. The high technology data collection methods such as magnetic tape memory, chips, and telemetering allow faster data processing by electronic equipment. A drawback to some magnetic tape and solid state data storage devices is that the person maintaining the meter often is unable to "read" the stored data in the field and make positive interpretations about how the meter is operating.

Circular charts are well suited for onsite cursory evaluation of data. All data can be viewed by an experienced technician in the field when the meter is maintained. A limitation when using charts is the need for manual data compilation, which can be time-consuming and is subject to human error. Short duration metering programs frequently dictate the use of chart recorders unless electronic data storage devices can be read and feedback given to field technicians on a timely basis.

The impact of bypasses or overflows must be assessed. Such discharges may occur only during wet weather and often require monitoring to avoid possible miscalculation of flow volumes during a sewer evaluation metering program.

It is usually desirable to subdivide a sewer system into subsystems to evaluate base flow, infiltration, and inflow in each subsystem. Without bypasses and relief sewers, division of a sewer system into subsystems is usually simple, based on direction of flow. Field work required for subsystem delineation should, where practical, be coordinated with flow meter reconnaissance inspections.

It is also important to recognize the limitation of temporary flow monitoring data. The levels of accuracy achieved at temporary open channel sites are seldom better than 10%. Metering inaccuracies tend to multiply and overlap in large flow monitoring projects that have many metering sites and require multiple subtractions and additions to define flow within individual districts. Moreover, accurate measurement of low flows from small areas often requires construction of costly weirs or other primary control devices. Where possible, individual meters provide for flow data from specific areas to eliminate addition or subtraction of flow. This procedure will significantly improve data reliability.

All flow-metering programs should be preceded by a thorough field reconnaissance to define the hydraulic characteristics of a proposed monitoring site (usually manholes) and to select alternative sites if unfavorable conditions exist at a preliminary selection site. A desirable open channel metering manhole has no change in grade or direction, a well-defined regular channel, moderate flow velocity to maintain self-cleaning, and good access. The flow conditions at a selected site will provide information needed to determine the best-suited equipment.

Field reconnaissance of metering sites should include: determination of pipe shape and size; measurement of sediment and flow depths; determination of accessibility; and sketch and description of manhole location. Also, an assessment of surcharge potential is very important. Consistent surcharging during wet weather may necessitate the use of pairs of leveled depth recorders or flow velocity and depth-recording equipment, because non-uniform flow or backflow conditions may occur during surcharging.

Existing lift stations can also be used to monitor flow. Running times for constant-speed pumps can be recorded with strip-chart recorders or digital recorders. The running time information can be converted to meaningful flow data if good pump capacity information is available. For pump station monitoring reconnaissance, a determination of the number of pumps and their operational status is needed. A determination of pump motor voltage may be required before certain event recorders are ordered. Ready access to the pump station wet well may be needed to perform drawdown capacity tests.

The installation of primary control devices such as weirs or flumes generally will increase the accuracy of the flow metering program. This is particularly true when unfavorable hydraulic conditions cannot be avoided. Weirs are probably the most accurate control section for low flow conditions. Solids deposition and buildup of rags on or behind a weir installation, however, can cause problems unless the backwater area is cleaned regularly. Flumes present less of a maintenance problem but their applications must be limited to sites where manhole clearance is sufficient and where excess approach velocities do not exist. Because control devices tend to reduce a sewer's hydraulic capacity, the installation of such structures should be coordinated with sewer system maintenance personnel.

Upon completion of a field reconnaissance for flow metering, the information to make intelligent equipment selections should be available. Metering equipment can be leased or purchased. Also, a number of companies provide wastewater flow metering services for sewer evaluation surveys.

Meter maintenance. Temporary wastewater metering locations are subject to a wide variety of potential pitfalls. A weekly minimum maintenance and inspection is recommended for temporary monitoring sites. A maintenance routine should include as a minimum the manual measurement of liquid level and a cross check of measured value against recorded flow depth. Other maintenance procedures depend on the equipment utilized. It is advisable to perform velocity measurements with the manual flow measurements at open channel sites to check equipment that automatically integrates area and velocity. A field log with notations for each visit to a site is recommended. Date, time, manually measured flow, and recorded flow should be logged along with any velocity data.

Data evaluation. Flow data evaluation, as required for a sewer evaluation survey

EVALUATION/REHABILITATION

project, usually entails producing flow parameters and hydrographs for each metering location. Data can be analyzed manually or automatically. The benefits of applying computer technology to flow data analysis include increased accuracy and production from the use of peripherals such as plotters, printers, and digitizers, and tape readers. Although data processing has definite advantages on large projects, analysis by manual methods will provide equally good results with the added advantage of the opportunity to individually access special conditions that may not be decipherable by a computer.

Normally, data are analyzed on 15-minute, 30-minute, or 60-minute intervals. Flow rates for each interval period are determined and a total daily volume is computed. Peak hourly rates can be determined by examination of individual hourly data. Hydrographs are produced on a similar basis with each data point falling at the appropriate interval. Figure 2.1 shows a typical weekly hydrograph for a monitoring site, with hourly rainfall data superimposed.

Precipitation measurement

Purpose and equipment. The measurement of precipitation as a part of sewer system evaluation is undertaken to correlate rainfall with flow metering data to determine the amount of inflow entering the sanitary sewer system. Several items are generally of interest: rainfall intensity, total volume per event, and duration of the event. Tipping bucket or continuous weighing rain gauges will provide all of this information. Charts are available for several durations. With some recorders, totalizers are available and provide a check against recorded data. In colder climates snow melting devices are also available on some models. For less sophisticated information, a manually read, graduated cylinder that records daily rainfall may be appropriate.

Sources of data. Before setting up a precipitation measurement system, other available data should be evaluated. Sources of precipitation data include the National Oceanic and Atmospheric Administration (NOAA), airports, state weather observers,

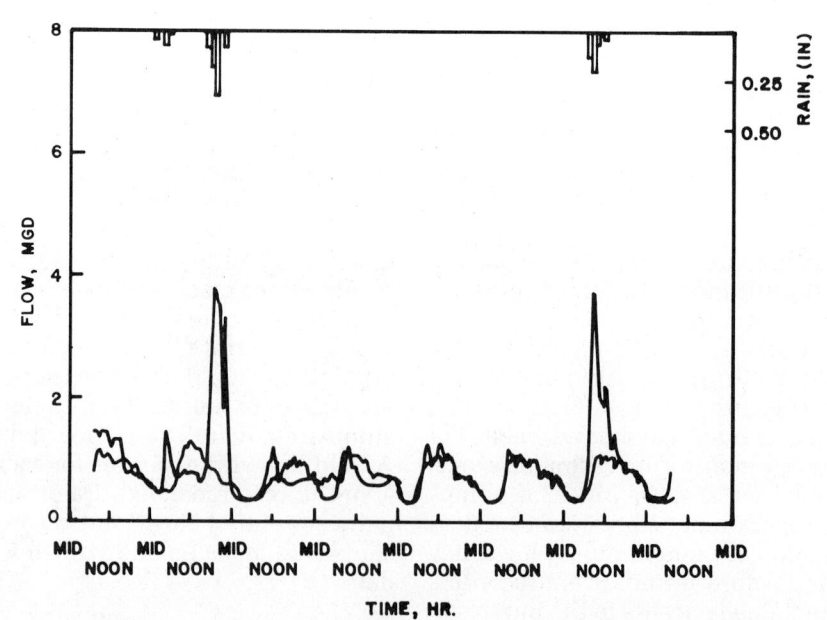

FIGURE 2.1. Typical weekly hydrograph for a monitoring site, South St. Paul (Dry Weather Flow, 5/29/81–6/5/81; Wet Weather Flow, 5/22/81–5/29/81).

EVALUATION METHODS

electronic media weather observers, other public works and research agencies, and private citizens. Admittedly, not all of these sources will have rainfall intensity data but they may help determine the distribution of rainfall after an event. NOAA has an extensive nationwide network of recording rain gauges. Those gauges with hourly rainfall data are summarized by state in a monthly publication entitled "Hourly Precipitation Data." Another useful publication containing daily precipitation quantities from all NOAA stations is "Climatological Data," also published monthly for each state.

Groundwater gauging

Purpose and equipment. Groundwater gauging is undertaken to locate the level of groundwater in soils and to indicate the variation in this level. Two types of gauges are commonly used for sewer evaluation studies: the manhole gauge and the piezometer. The manhole gauge, shown in Figure 2.2, is used to determine groundwater levels adjacent to the manhole. The gauges are inexpensive and fairly easy to install; however, they do clog easily from mineral deposits. The piezometer, shown in Figure 2.3, is generally installed in a hole excavated by a powered flight auger. Piezometers are more permanent and are far less prone to clogging. They also are more expensive, but with proper maintenance they should last for years and provide higher quality data.

Siting and data evaluation. Sites for piezometers should be away from under-

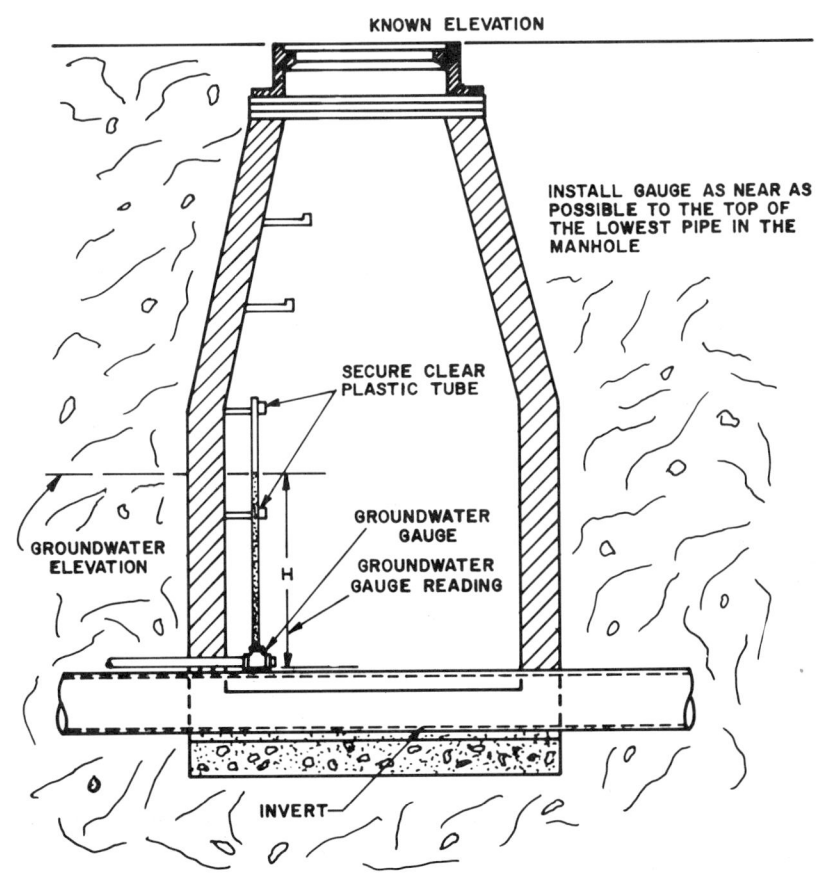

FIGURE 2.2. Static groundwater gauge installation elevation.

EVALUATION/REHABILITATION

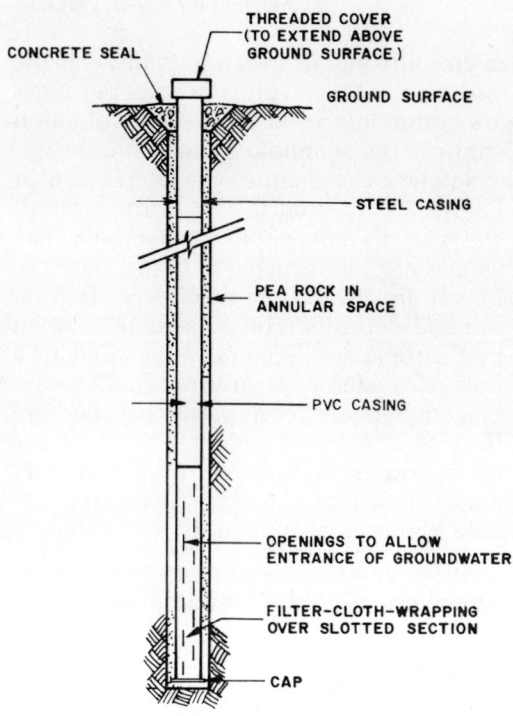

FIGURE 2.3. Groundwater gauge installation detail.

ground utilities and off streets far enough to prevent damage from street maintenance equipment. When locating these devices in parks or other areas that will be mowed, they should be staked or placed so that mower damage is unlikely. All caps should be threaded and locked if possible to prevent vandalism. Groundwater levels are then recorded on a periodic basis, perhaps weekly during flow-metering periods and biweekly or monthly thereafter. A plot of groundwater levels versus time is helpful in interpreting meter data and determining the levels of infiltration. In addition, certain field activities such as flow isolation measurements or television inspection can be scheduled with more confidence using groundwater elevation data.

Smoke testing. Smoke testing is a relatively inexpensive and quick method of detecting I/I sources in sewer systems. The method is best used to detect inflow such as storm sewer cross connections and point source inflow leaks in drainage paths or ponding areas, roof leaders, cellar, yard and area drains, fountain drains, abandoned building sewers, and faulty service connections. Smoke testing can also be used to detect overflow points in the sewer systems if groundwater is below the sewer.

If reliable information is to be derived from smoke testing, the method should not be applied to the sewer lines that are suspected of having sags or water traps. Either of these two pipe conditions may prevent the smoke from passing through, causing false conclusions. Similarly, the method should not be applied to sewer sections that are flowing full. Smoke testing cannot be utilized to detect the structural damages and leaking joints in buried sewers and service connections when the soil surrounding and above the pipes is saturated, frozen or snow covered. In each case, the smoke will be trapped and will not permeate through the surrounding soil even though there are cracks or leaking joints in the pipes. Thus, rainy and snowy days are not suitable for smoke testing. Also, the test should be closely monitored on windy days when the smoke coming out of the ground may be blown away so quickly as to escape accurate detection.

Positive findings during the tests pinpoint the I/I sources. Negative findings, however, do not necessarily prove that I/I does not exist.

The smoke bombs or canisters are used to generate the smoke required for the test. The smoke should be nontoxic, odorless, and non-staining. The 3-minute and 5-minute bombs are normally used, although there are bombs that last longer or for a shorter period of time. The air blower is used to force the smoke into the sewer pipes; a gasoline-driven blower is most convenient for this purpose. The air blower should have a minimum capacity of about 1500 L/s. The camera is used to take pictures of smoke coming out of the ground, catch basins, pipes, and other sources during the test. The photographs are taken for permanent documentation of the results. The sand bags and/or plugs can be used to block the sewer sections to prevent the

EVALUATION METHODS

smoke from escaping through the manholes and adjacent sewer pipes.

The police and fire departments should be notified daily of the test locations. Also media such as radio, newspaper and TV may be notified to inform residents of tests; in addition, residents should be informed individually on the day of testing. Person-

ATTENTION
Smoke Testing & Sewer Survey

(City Name)

For the next few weeks, inspection crews will be conducting a physical survey of the _____ sanitary sewer system. This study will involve the opening and entering of manholes in the streets and easements. An important task of the survey will be the "SMOKE TESTING" of sewer lines to locate breaks and defects in the sewer system. The smoke that you see coming from the vent stacks on houses or holes in the ground is NON-TOXIC, HARMLESS, HAS NO ODOR, AND CREATES NO FIRE HAZARD. <u>The smoke should not enter your home unless you have defective plumbing or dried up drain traps.</u> If this occurs, you should consult your licensed plumber. In any event, if the harmless smoke can enter through faulty plumbing, the potential exists for dangerous sewer gases to enter your home. Should smoke enter your home, you may contact a member of the smoke testing crew working in the area and he will be pleased to check with you as to where and why the smoke has entered your home. If you have any seldom used drains, please pour water in the drain to fill the trap, which will prevent sewer gases or odors from entering the building.

Some sewer lines and manholes are located on the backyard easement property line. Whenever these lines require investigation, members of the inspection crews will need access to the easements for the sewer lines and manholes. Homeowners do not need to be home and the workmen will not need to enter your house.

Photographs are to be made of leaks occurring in the system. We anticipate the smoke testing will require approximately _____ weeks in your area. Your cooperation will be appreciated. The information gained from this study will be used to improve your sewer services and may reduce the eventual cost to taxpayers.

THANK YOU.

Name
Address
Phone No.

FIGURE 2.4.   Sample smoke testing notice.

EVALUATION/REHABILITATION

nel should have proper identification. A sample written smoke-testing notice is shown in Figure 2.4, for additional notification, if necessary.

Smoke testing procedure
1. Isolate line sections to be tested. Up to three reaches or approximately 300 m (1000 lin ft) can be tested at a time, with or without plugging, depending on the conditions. Smoke testing without plugging has proved successful.

   Note any surcharged line sections at this time. Smoke will not pass through a flooded section. Extra set-ups may be required in this case.
2. Prepare basic smoke sketch, including
   - Location;
   - By: crew chief initials;
   - Date;
3. Smoke testing. The following points should be considered during any smoke testing program: use enough bombs to ensure complete smoke travels throughout the test section. Larger diameter sewers may require more bombs, or shorter test sections. Site blower near center of test section, if possible. Smoke should be generated continuously while visual inspection and photography is in progress.
4. Visual inspection. Walk entire area, front and back yards, and around buildings. Watch for smoke leaks from any source. Typical sources are roof leaders, driveway drains, house foundations, holes in ground over sewers or services, and storm sewer inlets. NOTE: Every house should have a sewer vent. This normally should smoke heavily. Roof vents should not be considered as smoke leaks.

   Photograph all leaks discovered, even if they cannot take in any water.

   Show leak location on sketch. Include the following:
   - Photo number and directions taken;
   - Description of leak, including address (or house number on sketch);
   - Provide ties to the leak;
   - Area and type of surface drained by leak.

Photographs. The photograph should show the maximum amount of smoke from the leak, and it should be close enough to show the exact source of smoke.

The photograph should be taken from far back enough to include the location of the smoke. A photograph of nothing but grass lawn is little help in finding the leak location at a later time. The same photograph showing some of the house would be better.

Photographs should be numbered in some logical consecutive manner to ensure the leak can be identified at a later date.

Manhole and pipeline visual inspection. The visual inspection of manholes and pipelines in the sewer system provides additional information concerning the accuracy of system mapping, the presence and degree of infiltration and inflow problems, and the general physical condition of the system.

Location techniques. Prior to any underground inspections, the manholes, or entry points, must be located. There is no substitute for clear, concise and up-to-date maps. These usually can be obtained from the owner's maintenance or engineering group. The accuracy of any maps used should be confirmed with operating personnel.

If accurate maps are unavailable, a preliminary working set must be prepared using all existing plans, reports, and as-built drawings; discussions with operating and engineering personnel familiar with the system are also important. Another source of data would be former operating personnel and engineers, or long term residents. The interviews must be directed to strategic areas of concern such as the location of demolished buildings, and the rerouting of streets. All information must be evaluated, and pertinent data should be used in the preparation of the preliminary working maps.

With the preliminary system mapping completed, the actual field work can commence. From known locations, the system can be traced by opening manholes, iden-

EVALUATION METHODS

tifying incoming and outgoing lines, establishing line-of-sight, and pacing off to the next manhole using distances as recorded on the maps. This is, of course, complicated in many cases by building changes, street rerouting and paving, and landscaping. The use of metal detectors, probe rods, transits, measuring wheels or tapes, dye tracing, and smoke testing can be used in solving many problems encountered. Starting in upper areas of drainage basins and following the flows from there is usually productive. Sewers near drainage basin boundaries generally are laid in much shallower trenches following the natural drainage of the ground surface. A key point to remmber is that by identifying the conditions governing the installation of the pipeline, it is much easier to locate that pipeline.

The location of manholes should be recorded on the maps, with any changes noted; if further work might be performed at that location, it is useful to paint the manhole's identifying number on the cover.

Inspection procedures. The actual inspection of manholes and pipelines depends upon the type of information required. For instance, if the area's problems have been determined to be inflow only, a surface investigation may be all that is required to assess the conditions properly. If, on the other hand, groundwater leaking into pipes through deteriorating joints is suspected, it may be desirable to inspect the pipes visually by using a closed-circuit TV camera or, if large enough, by actually walking the line. If the presence of infiltration is being investigated, the inspections should be performed during a period of high groundwater.

Data that can be obtained from the surface include:

- Exact location of the manhole;
- Diameter of clear opening of the manhole;
- Condition of the cover and frame;
- Any other defects to the cover and frame that would allow inflow to enter the system;
- Whether cover is liable to be under water;
- The potential drainage area tributary to the defects;
- Type of material and condition of the corbel and walls;
- Condition of steps; and
- Configuration of the incoming and outgoing lines.

By entering the manhole and using equipment such as portable lamps, mirrors, rulers, and probe rods, the following data can be obtained in addition to that identified from the surface:

- Type of material and condition of apron and trough;
- Any observed infiltration sources and the rate of infiltration;
- Indications of height of surcharge;
- Size and type of all incoming and outgoing lines; and
- Depth of flow indications of deposition and the characteristics of flow within all pipes;

By viewing the incoming and outgoing lines with a mirror it is possible to ascertain the following:

- Structural condition;
- Presence of roots;
- Condition of joints;
- Depth of debris in lines;
- Depth and approximate velocity of flow; and
- Location and estimated rate of any observed infiltration.

The safety precautions when performing manhole or pipeline inspections start when the vehicle arrives at the work site. Any time traffic is to be disrupted, the appropriate authorities must be notified. Many municipalities require permits to be issued before any diversion of traffic can occur. If possible, work should be scheduled to avoid rush hour traffic. Warning signs and flagbearers must be placed far enough in front of the work area to allow motorists to slow down for the work area. Evenly spaced traffic cones should be used to channel traffic around the work area. Placement of work vehicles between on-

EVALUATION/REHABILITATION

coming traffic and the workers is an effective shield in most cases. Reference material on this subject is available.[2]

There are six major categories of hazards encountered in entering manholes. These hazards are presented in descending order of known frequency of accidents and deaths to sewer workers.

1. **Atmospheric hazards** consist of three major types—explosive, toxic, and lack of breathable oxygen. Unpleasant odors are usually present in the sewer but many are only dangerous because they shift attention from other lethal conditions.

   Explosive or flammable gases can develop at any time in the collection system. Methane is a product of biological decomposition prevalent in stagnant conditions caused by restricted flow. Even though methane is lighter than air, the danger is not limited to the arch and upper manholes of a system. Propane, gasoline, and many flammable solvents that find their way into the sewer are heavier than air and tend to form pockets in the lower reaches of the collection system. Care should be exercised in all sections of a line.

   Toxic conditions are most likely a result of hydrogen sulfide formation. Hydrogen sulfide is produced by the decomposition of materials containing sulfur. Because hydrogen sulfide is heavier than air, it will accumulate in the lower sections of a collection system. Hydrogen sulfide is easily detectable in low concentrations because it smells like rotten eggs. The sense of smell, however, should not be the only method of checking for hydrogen sulfide. Because hydrogen sulfide is detectable by odor only in low levels of concentrations, never assume the condition does not exist if the smell cannot be detected, because high concentrations tend to deaden the olfactory senses.

   The amount of breathable oxygen in a manhole may be reduced or even eliminated entirely if the air is replaced by another heavier-than-air gas. If there is no breathable oxygen in a manhole, the life expectancy of anyone entering the manhole is approximately 180 seconds.

2. **Physical injury** during manhole inspection can occur from several causes. The manipulation of tools in restricted spaces with unsure footing often results in bruises and strained muscles. Corroded steps lead to falls and cuts. Dropping tools to workers and throwing them out again can cause eye injuries and facial cuts. A secondary hazard from any injury inside a manhole is the ever-present danger of infection from the unsanitary environment.

3. **Infections** are always a risk when entering a manhole. Every disease, parasite, and bacteria of a community can end up in the wastewater collection system. Personal cleanliness is the best means of protection.

4. **Insects and animals,** although less dangerous to workers than infections and diseases, can be a hazard. Always inspect a manhole for insects, bugs, rodents, and snakes before entering.

5. **Exposure to toxic acids, bases, and other hazardous liquid or solid chemicals** that can be discharged into the wastewater collection system either by accidental spills or deliberate action by industry or the public is always a potential health hazard. Proper boots and gloves are effective means of protecting against these toxicants.

6. **The chance of drowning** while working in a manhole is increasing with the construction of more and bigger interceptor pipelines. Always wear harnesses with lifelines attached before entering any manhole.

Those employees who are to enter a manhole should be in good health, should have no open sores or skin irritations, should not be under the influence of alcohol or drugs, and should have immunizations up to date.

Where insects are a problem, spraying with an insecticide is suggested. If a manhole that is unusually filthy or highly odorous must be entered, the walls and apron can be washed down with a high velocity stream of clear water an hour or so before entry.

The following should be considered when entering a manhole:

- Never use only hands to open a manhole. The cover should be opened with a hook and dragged away from the manhole frame.
- Manholes upstream and downstream should be opened to encourage natural ventilation. A blower should be used to ventilate the manhole with fresh air. If the blower is gasoline powered, the exhaust must be downwind of the manhole.
- The area immediately around the manhole, including the ring, should be swept clean of any loose debris or dirt.
- A gas detector should be employed the entire time that a worker is in the hole.
- The person entering the manhole should wear a hard hat, gloves, steel-toed shoes, long-sleeved coveralls, and a safety harness.
- The lifeline should be held continuously by a crew member above ground. This lifeline holder's only job should be watching the person in the hole at all times. Tying the lifeline to an object is a dangerous practice. If a passing vehicle were to hit the lifeline or the object to which it was tied, the worker in the hole would be seriously injured. Upon completion of the task, the lifeline holder should assist the exit of the manhole worker.

Data recording. As discussed above, the data to be obtained from each inspection should be tailored to the objectives of the particular project. Experience has proved that checklist-oriented recording forms are more appropriate than those requiring a long narrative. It is important, however, that the forms include an area to allow the recording of notes about conditions observed during the actual inspection. In addition to a written log, the inspection should be documented by either still photographs or videotapes of the entire inspection with a verbal description of each defect entered into the audio portion of the tape. This procedure may be modified to tape only defects as they are located and not the actual progress of the inspection, depending upon the documentation requirements.

Dye-water flooding. Dye-water flooding is one of the rainfall simulation techniques used to identify specific defects that can contribute I/I during rainfall or snowmelt. Additionally, dye-water flooding can be effective in quantifying the amount of I/I that can enter a section of sewer or specific defect under a controlled runoff situation.

The following are situations where dye-water flooding may be considered:

- Storm drains that parallel or cross sanitary sewer sections (including services) and have crown elevations higher than the invert elevations of the sanitary sewers.
- Stream sections, drainage ditches, and ponding areas located near or above sanitary sewer lines (including services).
- Yard, area, and foundation drains, roof drains, abandoned building sewers, and faulty and/or illegal connections.
- Verification and/or quantification of actual or suspected problems identified from other tests such as smoke testing or physical survey.
- As color cameras become more common in sewer line televising, dye-water flooding may be utilized to determine if a leak observed on the camera is a result of runoff.

The equipment needed for the dye-water testing is limited to that required to carry the water to the testing site and to block the sewers or the study areas before the testing. When fire hydrants are close to the sewer sections to be tested, a fire hose is all that is needed to deliver the water to the testing site. On the other hand, when the water source is not close by, water tankers will be required to deliver the water. Sand bags or sewer pipe plugs normally are used to block the sewer sections.

Fluorescent dyes usually are used for the test. Each dye has a distinct color that is readily detectable by eye. A suitable dye should be safe to handle, visible in low concentrations, miscible in water, inert to the soils and debris in the sewers, and biodegradable.

Depending on the infiltration and inflow sources to be identified, and the configuration of the runoff situation that is being simulated (that is, storm drain, drainage ditch, spot flood), the procedures for dye-water flooding differ. Five examples are provided below.

1. Determininination of I/I conditions caused by storm sewer sections. Storm drains that parallel or cross sanitary sewer sections and have crown elevations greater than the invert elevations of the sanitary sewers can be sources of rainfall induced infiltration or inflow. They are inflow sources if there are cross-connections between the storm drain sections and sanitary sewers. They are infiltration sources if the storm water can exfiltrate from them, percolate through the soil, and enter the sanitary sewers through pipe defects, broken pipes, or leaking joints. The general procedures for dye-water testing in storm drain sections are as follows:

- Plug both ends of the storm drain section to be tested with sand bags or other materials. Block all the overflow and bypass points in the sewer section. Provide bypassing of flow, if necessary.
- Fill the storm drain section with water from fire hydrants or other nearby water sources. Add dye to the water.
- Monitor the downstream manhole of the sanitary sewer system for evidence of dyed water.
- Measure the flows in the manhole before and during the dye-water testing. As an alternative, the flows can be simultaneously measured at both the upstream and downstream manholes during the test.
- Record the location of storm drains and sanitary sewer lines being tested, the time and duration of tests, the manholes where the flows are monitored and the flow rates, the observed presence, concentrations and travel time of the dyed water into the flow monitoring manholes, and the soil characteristics.

2. Determination of I/I conditions caused by stream section. To determine whether the stream sections, ditch sections, and ponding areas located near or above sanitary sewer sections are causing I/I conditions in the sanitary sewers, a procedure similar to that described above is recommended. In these cases, the stream sections, ditch sections, and pond areas to be tested should be plugged or dammed (if necessary) and filled with dyed water to the desired levels. The presence, concentration, and travel time of the dyed water into the sanitary sewers are then monitored in the downstream manholes. Weir measurements or depth and velocity measurements should be made where quantification of I/I is desired.

3. Identification of roof leader, cellar, yard and area drains, abandoned building sewers, faulty connections, and illegal connections. Most of these inflow sources are located on private properties. The property owners should be notified before the tests to identify the aforementioned inflow sources. To identify the above-mentioned inflow sources, dyed water is poured into the corresponding fixtures and their presence is checked in the closest downstream manhole in the sanitary sewer system. The date of the test, the address where the inflow sources are identified and the type of inflow sources should all be recorded. Again, weir or depth and velocity measurements can be made at the downstream manhole to quantify the source.

4. Identification of structurally damaged manholes. The dye-water test can also be used to identify the structurally damaged manholes that impose potential I/I problems. This is accomplished by flooding the area close to the suspected manholes with dyed water and observing the presence of the dyed water at the manhole walls.

5. Verification and/or quantification of Actual or suspected I/I sources found in other field investigation phases. The dye-water test can be effective in verifying suspected sources of I/I identified in a physical survey or smoke testing study. Quantification of the defects can be done at the same time. The log sheet from the field study is used to identify and locate the source if the area of the suspected source is flooded. In some cases it may be necessary to restrict the runoff with sand bags to allow the area to become saturated. The downstream manhole is monitored for presence of the dyed water. If a positive result occurs, a weir or depth and velocity measurement is taken to quantify the source. This works well for findings such as: manholes subject to surface runoff; holes in ground smoking over services or main lines; large areas of ground smoking over or near services or main lines; and cracks in street pavement smoking.

Safety measures are discussed below.

• A gate valve should always be used to control the flow of water from a hydrant. Hydrants should always be opened and closed slowly and completely. Never use the hydrant to throttle the flow; this may damage the hydrant. Use a pressure gauge when inflating plugs or bags. Never over-inflate a bag or plug; damage to the pipe may occur or it may explode and injure the technician.

• Always attach a valved air line to a bag or plug so that it may be deflated from the surface. Water head build-up behind a bag or plug in a sewer line is powerful and dangerous. If a bag or plug lets loose, the technician may be seriously injured. He or she may also be pinned against the manhole wall and possibly drown if enough water were backed up. The larger plugs present the most dangerous situations.

• Always watch the water level and be aware of the minimum level that will cause flow to back up into buildings and cause property damage.

• Provide proper traffic control where hoses cross streets. Damage to both the water system and cars can result from hitting the hoses at high speeds.

• Always secure the discharge end of the hose. High pressure and flow can cause the end to go into a wild whipping action, injuring personnel and damaging property.

• Remove all plugs when a setup is complete. Failure to do this will result in backup and property damage.

Dye-water flooding is limited to locations where large quantities of water are available for the test. It is usually not practical to flood an area that is more than 150 m (500 ft) from a fire hydrant because of the amount of equipment necessary and setup time. Where hydrant water is not available, tank trucks can be used. Tank trucks, however, are limited in both the rate and total quantity of water that can be applied and, there is considerable time lost in refilling the truck.

Both sanitary sewer and storm drain manholes may have to be entered for the test. Water head buildup behind plugs is dangerous. Caution and safety are extremely important. In some places water use may be restricted by either rate of use or total quantity available.

Once a positive dye transfer has occurred on a line segment, it may not be possible to conduct another test on that line or downstream for some period of time because of the presence of the dye. This is most limiting when more than one spot check is desired on a line segment.

Once a positive dye transference is observed it often is necessary to inspect the line internally so that the actual location of the leak and proper rehabilitation can be determined. It may be desirable to reflood the setup at the time of televising to identify the defect positively.

A field log sheet should be filled out for each dye-water test whether or not a positive transference is observed. The following data should be included on the log sheet:

• Date;
• Time;

EVALUATION/REHABILITATION

- Field Crew;
- Location;
- Type of setup;
- Sketch of the setup;
- Sanitary manholes checked;
- Dye transference information;
- Flow readings before transference and during transference;
- Flooding time;
- Pipe size and storm footage involved;
- Comments on the setup.

The sketch should be very clear. It should indicate exactly what was flooded and the relationship between that and the sanitary sewer system. It is often desirable to photograph the setup. The photograph number should be recorded on the sketch for reference.

The comments area on the log sheet can be used to record observations such as: previous and existing weather conditions, soil conditions, access for future maintenance or rehabilitation, unusual conditions in the storm or sanitary systems, and difficulties incurred in performing the test.

Night flow isolation. Night flow isolation is a technique employed to determine the amount of extraneous water, usually infiltration, entering a reach of sewer. It differs from other flow measurements in that the primary purpose of performing night flow isolation is to determine the specific reaches of sewer that have excessive infiltration, so that further action, usually internal inspection, may then be performed.

For economical and reliable planning and execution of a flow isolation program, an accurate map of the system is highly desirable. The proper selection of measurement points is critical to the success of an isolation program, and the map must accurately reflect the layout of the stream for proper selection of these points. Also, the efficiency of field execution is often drastically affected by the reliability of the maps supplied. The map should contain information on all pipe diameters and lengths, and permit ready location of the manholes.

Additional information beneficial to performing the night flow isolation technique include:

- Manhole accessibility;
- Estimate of flowrates;
- Cleanliness of pipes;
- Sizes of weirs and plugs required;
- Large users;
- Night users.

The ultimate usefulness, accuracy, and cost of flow isolation is strongly dependent upon the selection of the proper length of reach to be isolated. The two major determining factors are measurement accuracy versus reach length, and the effects of possible groundwater migration.

Measurement accuracy versus reach length. Regardless of the flow measurement technique utilized, the measurement will always have some component of error, which may be in proportion to the flow rate being measured. The sources of these errors are associated with both inherent limitations of the techniques and limitations that arise in practical applications. For example, an inherent limitation of a weir is the accuracy of the assigned coefficient of discharge that enters into the weir calibration; a limitation that arises in practical application is the accuracy with which it is possible to determine the water level over the notch while reading the weir in the field. Each measurement technique has its own sources of error. In general, an inaccuracy of up to 10% of flow rate can be expected of a properly executed flow measurement.

If the measured flow during flow isolation is purely infiltration, this error is not significant. Difficulties arise, however, when non-infiltration flow also is present. This is more evident when employing a differential of two measurements to obtain a net measurement, as in differential isolation.

Two alternatives are available to minimize this type of problem:

- Reduce the total flow rate to be measured. This can generally be done only by plugging and is useful for moderate to small flow rates.

- Increase the amount of infiltration to be measured. This can be done by increasing the net length of reach under measurement. For this to be effective, the sanitary component of the flow rate generated within the reach must be minimized.

For both alternatives, it is clearly an advantage to have the sanitary flow component of the measured flow as small as possible. For this reason, such measurements are generally conducted during the minimum flow, typically from midnight to 6 a.m.

Thus, differential measurement techniques generally are not suitable for infiltration isolation of very short reaches.

Groundwater migration. The selection of reach length to be isolated is dependent on the possible effects of groundwater migration. In most circumstances, infiltration into a pipe is not confined to singular sources, but rather is distributed, with many potential entry points for any given groundwater condition. Often, groundwater excluded from one section of pipe will migrate along the pipe trench and enter at other defects. For this reason, it is impractical to consider rehabilitation of defects on a point-by-point basis. If any rehabilitation of a pipe is attempted, all possible points of entry should be repaired and considered in the I/I correction and cost estimate.

Isolation of infiltration down to small reaches is therefore unnecessary in most circumstances. The length of reach to be isolated should reflect the length that is thought to have a common infiltration problem and should also reflect the length of reach that will be considered for rehabilitation as a unit.

Other considerations in the selection of measurement points, and thus reach lengths to be isolated, are access, suitability for measurement, the costs of obtaining the measurements, and the potential costs of internal inspection and ultimate rehabilitation. Selecting longer lengths of reach will result in fewer measurements and a less costly isolation program. The selection of longer reach lengths, however, also will result in higher total costs for internal inspection and rehabilitation. The latter is somewhat mitigated by the fact that the cost per unit length of internal inspection and rehabilitation is lower for longer continuous runs of pipe than it is for the same length of shorter, discontinuous runs because of money saved in access and set up costs to perform the work.

Program design. Single-pass method—This method of isolation consists of prior selection of all measurement points from map studies and then execution of all field measurements. Its advantage is simplicity of planning and execution. The disadvantage of this method is that it is inflexible and thus the planned measurement points may be unsuitable or unnecessary in some circumstances.

Multi-pass method—This method of isolation involves selecting a few key measurement points for field execution and then analyzing the results prior to the selection of further measurement points in areas of interest. It is more complex but allows concentration of effort in high infiltration areas.

Both methods are applicable; circumstances should dictate the choice. If infiltration is suspected to be generally distributed, the single-pass method is more useful and economical. Conversely, if infiltration is confined to smaller areas, the multi-pass method may be less costly.

Isolation techniques. The objective of the flow isolation program is to isolate small reaches of sewer and measure the infiltration rate within each of these reaches. Two methods are available to isolate specific lengths of sewer: plugging and differential isolation.

Plugging—This method consists of physically isolating the sewer length from the rest of the system by means of plugs inserted into the sewer pipes. Plugs are generally pneumatic and are available from several manufacturers in diameters up to 600 mm (24 in.) for rigid body plugs and may be available up to 1000 mm (42 in.) diameter for non-rigid bag plugs.

EVALUATION/REHABILITATION

Plugs need to be inflated; thus compressed air must be available at the site. Also, the installation of larger diameter plugs may require winching into place. Plugs generally are inserted into an incoming line for easier deflation.

It is very important to note that plugs may be subject to large total hydraulic and pneumatic forces even at low heads and small diameters. Failure of plugs is not uncommon; workers near or downstream of plugs must exercise extreme caution at all times. All plugs should be tied off to manhole steps and have a stout tag line carried to the manhole entrance. The tag line is used to recover blown plugs and remove deflated plugs.

Plug removal is the most critical part of the plugging operation. Deflation is best done from above grade by use of an extension hose on the plug valve. If the manhole must be entered to deflate a plug, the person entering should wear a safety harness and never should stand in front of the plug. The plug should be deflated gradually. Even with care, a plug may be ejected from the pipe with high velocity, and the manhole may rapidly fill with the released water. The person in the manhole should be positioned in such a way as to avoid the plug and exit the manhole as rapidly as possible.

After a section of sewer is isolated by plugging, sufficient time must be allowed for the pipe to drain down before any measurements are attempted. This time should increase with the length of sewer isolated and the average slope of the sewer. The draindown time may be estimated roughly by assuming a flow velocity of approximately 0.15 m/s (0.5 ft/sec) from the plug to the point of downstream measurement. Before any measurements are taken, however, an equilibrium flow situation must be verified at the point of measurement by the method of successive readings described below.

Plugs inserted in the pipe will, of course, interrupt the existing flow and cause wastewater to be stored behind the plug. Prior to plugging, the flow rate, the volume of storage available, and the relative elevations of the pipe and local basements should be estimated to determine the safe length of time the plug may be left in the line. The relationship of the safe plugging time to the draindown time of the pipe will determine the maximum length of sewer that is feasible to isolate by plugging; however, the estimated safe plug time should never be relied on to avoid flooding of basements. Frequent visual observation of actual water depths in upstream manholes is vital during a plugging operation.

When a plug is removed, a surge of water may be propagated for some distance downstream. This may interfere with downstream measurements, may blow downstream plugs that are lightly seated, or may even flood lowlying downstream basements. For these reasons, plug removal should be as gradual as possible and properly sequenced with other operations conducted within the same sewer. The manufacturer's instructions on the proper use and maintenance of plugs should always be followed to avoid injury or property damage.

The advantages of plugging for flow isolation are that it physically isolates the length of sewer under consideration and reduces the flow rate required to be measured downstream. This tends to increase the accuracy of the flow measurements obtained, especially if obtained by use of portable weirs.

Differential isolation—This method involves subtraction of all flows coming into the section from all flows going out of the section to obtain the net increase of flow within the section itself. With this method, the flows are not physically interrupted, which eliminates the problems associated with the upstream storage of plugged flows.

To obtain a usable net flow rate using this method, sufficient care must be taken to ensure that upstream measurements do not have an adverse effect on downstream measurements. The entire system of flows and measurement devices associated with the isolation of a given section of sewer should be in equilibrium. Equilibrium is most important when using portable weirs for flow isolation, because a certain amount of time is needed to stabilize them after installation. This can be achieved first by setting all weirs and allowing the entire system to reach equilibrium before taking any readings, or by setting, reading, and removing the weirs.

Measurement techniques. Portable weir methods—A common measurement technique used in flow isolation is to insert a portable V-notch (or other) weir into the incoming pipe at a manhole. Portable weirs are commercially available to fit circular pipes. An advantage of portable weirs is relative ease of installation, and they generally are calibrated for direct reading of flow rates. For a usable measurement, care must be taken to ensure that:

- The weir is installed level;
- The weir is properly seated and watertight at its perimeter;
- The nappe is aerated;
- The velocity of approach is small;
- The flow over the weir has reached an equilibrium condition; and
- The weir is properly read.

Under ideal conditions, a V-notch weir may have an error proportionate to the flow rate plus an error due to flow depth estimation, which increases with low flow rates. Almost no real sewer system provides measurement points meeting all of the above conditions; therefore portable weir measurements must be considered at best an estimate of actual flow rates. Sufficient care should be taken to minimize adverse conditions; however, reasonable judgment must be used in interpretating results.

Despite the above limitations, portable weirs are considered the most practical and cost-effective method of obtaining flow isolation measurements in most circumstances.

Velocity-area method—Estimates of flow rates at a measurement point may be made using the velocity-area relationship:

$$Q = AV \quad (1)$$

where

Q = flow
A = cross-sectional area, and
V = mean velocity.

To utilize this relationship, both the cross sectional area of flow "A" and the mean velocity "V" must be determined.

The cross sectional area of flow can be obtained by first measuring the depth of flow at the point where the velocity reading is obtained. The depth of flow may be measured by pipe caliper, or less satisfactorily by ruler.

If sediment is present, the depth of debris must be estimated by seating a ruler on the pipe invert and marking the apparent top of the debris on the ruler by feel, withdrawing it from the flow to obtain the reading. The cross sectional area of flow "A" may be computed by reference to prepared tables or by geometry. The sediment area should be subtracted from the total wetted area to obtain the actual area of flow.

The mean velocity "V" of the flow may be obtained by using magnetic or propeller current meters, by interval timing, or by timing of a floating object. The use of magnetic or propeller current meters to measure open channel flow in pipes is described in the literature, and the manufacturer's directions should be consulted for each particular type of instrument.

Dye interval timing also is described in the literature and may give reasonably accurate results provided that:

- The velocity of flow is approximately 0.6 to 1.2 m/s (2 to 4 ft/s). Extremely slow flows may cause irregular dispersion of the dye slug or may make determination of the boundaries of the slug difficult; extremely fast flows may cause difficulty in the timing of the leading and trailing boundaries of the slug and may make insertion of a well defined slug of dye difficult.
- The pipe is clean, free of roots, uniformly graded and the flow depth is constant over the length of the pipe. Deviations in any of these factors will introduce inaccuracies into the technique. Without

EVALUATION/REHABILITATION

extensive examination, it may be impossible to determine if all such conditions are suitable for a given pipe. Furthermore, at the low flow depths typical of night flow isolation measurements, the effects of these factors are magnified.

Dye interval timing may be useful in flow isolation under certain circumstances, however, because of the above limitations it cannot be considered a generally applicable technique.

The use of floating objects to estimate mean velocity is subject to many variables and is not accurate enough to be of great value in flow isolation. The main item of importance in conducting flow isolation is a reasonably accurate map of the sewer system showing pipe and manhole locations, lengths, and diameters. Without such a map, it is virtually impossible to properly plan or execute night flow isolation, or interpret the results in a meaningful way.

Fluorometric methods—Fluorometry, also known as dye dilution, is a useful flow measurement technique in sanitary sewers, because it is independent of sewer dimensions, velocities, conditions, and surcharging. It is particularly useful in large diameter sewers. Dye is continuously injected upstream at a constant rate sufficiently far from the site of measurement(s) so that the dye is thoroughly mixed and its concentration is uniform throughout the cross section of wastewater at the point of sampling. Under these conditions, a rate of flow can be calculated anywhere downstream where dye is present by the following relationship:

$$Q_1 C_1 = Q_2 C_2 \quad (2)$$

where

Q_2 = the discharge rate sought downstream,
Q_1 = the rate at which the dye is injected in the same units as Q_2,
C_1 = the concentration of the injected dye, and
C_2 = the concentration of the dye at the point of measurement(s).

The dye's change in concentration at any point in the sewer is proportional to the change in flow rate. The proportionality is defined by the variables of the above equation. Of the four variables, three are known and used to calculate the fourth. The flow rate or discharge rate sought downstream, Q_2, is calculated by:

$$Q_2 = \frac{Q_1 \times C_1}{C_2}$$

Results. Expression—Flow isolation data are generally expressed in terms of (L/m.d)/m (gpd/in.diam/mile). The measured flow is divided by the pipe diameter and the isolated length.

Interpretation—Care must be used in drawing conclusions from the data. A high infiltration rate is not bad in and of itself. The rate must be evaluated both qualitatively and quantitatively. Other factors such as sewer capacity, projected future needs, treatment costs, and basement flooding problems must be considered. For example, many communities or sanitation districts impose even tighter requirements because of limited sewer capacity. Reducing infiltration creates additional sewer capacity for growth. Conversely, cities with adequate capacity and low treatment costs can cost-effectively justify higher allowable levels of infiltration. Each community should evaluate its own situation. The accuracy of the measurements is also a limiting factor.

Pipeline cleaning and television inspection

Pipeline cleaning needs and equipment. Pipeline or sewer cleaning is necessary for both efficient collection system operation and for improving the effectiveness of television inspection. Standards for cleaning that are acceptable for system operation will provide for adequate television inspection results. Standard system operation cleaning should be regarded, however, as a minimum acceptable effort in regard to television inspection.

As with any cleaning, all the collected

EVALUATION METHODS

sediment and debris should be removed from the line and disposed of at an approved site. Care should be taken during the cleaning to ensure that only minimal amounts of deposition are lost into downstream lines.

An extensive outline of sewer cleaning procedures can be found in WPCF manual of practice no. 7.[3]

Television equipment and procedures. Television inspection is accomplished by using closed circuit systems specifically designed for sewer inspection. There are several configurations of closed-circuit systems for sewer inspections, each of which have the following in common:

- Power for operation generated on site;
- Power control;
- Transport winches;
- Video (color, if possible) and lighting control;
- Recording and documentation; and
- Radio communication.

Each equipment format has unique advantages and disadvantages; however, systems that allow the TV van operator to control both the speed and travel of the camera and that allow for remote control of the camera itself are the most efficient and provide the highest quality pictures. Additionally, systems that use only one or two conductors to control all viewing functions are the most reliable and the easiest to troubleshoot.

Data recording. The effectiveness of television inspection is directly related to the completeness and accuracy of the collected data. For each pipe line or sewer inspected, records should be collected on a field form, videotape, or in photographs.

The inspection form should contain the following data for each manhole-to-manhole section that is inspected:

- The date of the inspection.
- The reason why the inspection was done.
- The location of the pipeline and the upstream and downstream manholes.
- The compass direction of the viewing and the direction of the camera's travel.
- The pipe size, type, pipe joint length, and the overall footage of the inspected sewer.
- The quantity of I/I expected to be observed, the actual quantity of I/I observed, and the total quantity of I/I measured in the sewer at the time of the inspection.
- A description of each service connection and defect observed and their distance from the point at which the viewing began.
- The observer's assessment of rehabilitation recommendations for the inspected sewer and cost estimates.
- Reference should be made on the log to each photograph taken and to the video tape of the entire inspection.

For each pipeline inspected, the videotape should contain the entire pipeline, regardless of its condition. This ensures that the work was done and completed in an efficient manner. It allows the engineer to double check recommendations, the accuracy of the work and the overall effectiveness of the entire TV program. It also provides a final check to ensure that no defects have been overlooked.

The videotape also should always include a brief and informative verbal description of the pipeline that is being inspected. The narration should contain no reference to cause and effect in regard to defects, and also should contain no recommendations on rehabilitation. Finally, the narration should contain absolutely no reference to any party's liability as it relates to defects, payments for rehabilitation, or personal opinion.

Photographs should be taken of each severe structural defect and all significant sources of I/I. Typical pictures of the pipe or the pipe joints are optional. Photographs are acceptable when there is an urgent need to illustrate defects to interested parties.

Leak quantification. The quantification of I/I sources during a sewer system evaluation survey is based on the type of source

25

EVALUATION/REHABILITATION

and physical characteristics of the sewer system and/or the area of the leak. In many cases, they are based on empirical values instead of straight engineering principles. Experience is a valuable tool either in determining the quantity of a source or obtaining the proper physical observations that enable a good estimate of how much I/I is leaking into the sewer system.

To determine the quantity of a leak it is first necessary to know whether the source is infiltration or inflow. They can be distinguished by the duration of their response to rainfall. Regardless of how leaks enter the system, those that last from a few days to several months are considered infiltration. Those flows that last a much shorter period are considered inflow. Inflow can be further broken down into direct and indirect sources. Direct inflow sources enter the system directly as a result of runoff and have a very short duration. Examples of direct inflow sources are point source ¼catch basins and roof leaders.

An indirect inflow source is one in which runoff has to pass through a medium before it enters the sewer system. Sources of this nature include pipe joint leaks, which occur because of transference from storm drain or an area of surface drainage, manhole wall leaks with the same conditions as the joint leaks, and foundation drains. Similar to an infiltration source, these sources have a much longer duration (throughout the duration of the rainfall event) and will contribute a more constant flow throughout a rain event. Direct inflow sources will contribute a much larger rate of water, but again only for a short period of time. Almost all direct inflow sources can be quantified by application of the Rational Formula; with a trained field technician obtaining physical measurements, good estimates can be determined. Indirect inflow sources are largely discovered during dye-water flooding. The most accurate method of determining the quantity of leaks in the pipe is first to install a portable weir into the line to be tested before dye-water flooding is undertaken, and then after the transference is completed, re-install the weir again to measure the increase in flow caused by the line defects in the sewer system. During subsequent television inspection, the already measured increase in flow will give a basis for the quantity of flow from these defects. In some cases it is not always possible to obtain a weir measurement because of the condition of the line, and it is only possible to obtain a depth measurement both before and after the increase in flow. To estimate this quantity, Manning's Equation for open channel flow could be used and is as follows:

$$Q = (k/n) R^{2/3} S^{1/2} A \qquad (3)$$

where

Q = the flow rate,
n = the Manning (Kutter) roughness factor,
R = the hydraulic radius,
S = the energy loss per unit length of pipe,
A = the area of the discharge, and
k = a conversion factor: 1.00 for metric and 1.486 for English units.

Probably the most difficult sources to quantify with any degree of accuracy are manhole lids subject to surface runoff. An approximation of the flow entering through lids is given in Table 2.1 and is based on a variety of manhole lids as to their ability to contribute inflow under different heads.[4]

When not enough data are available to allow the use of flow equations (for example, leaking manholes, small leaking joints), estimates must be made at the time of discovery. These estimates must be based on prior training and experience in judging sources. In some cases, leaks can be captured in a bucket of a known volume and timed as to how long it takes to fill. This is a very simple but effective method.

The accuracy of all I/I estimates is influenced by the empirical nature of the task. Study conditions do not always match the parameters of the flow equations used and in these cases engineering judgments are necessary to apply the equations. These judgments are based on past experience in

TABLE 2.1. Inflow through manhole frames and covers.

Manhole Frame & Cover Preparation	Surface Inflow Avg. (gpm)[2]		
	Test 1	Test 2	Test 3
Non-machine bearing surface	3.99	10.63	16.80
Machine bearing surface	1.06	1.81	2.54

[1]Based on test conditions of:
  1. Splashing water on lid simulating steady rainfall with no ponding;
  2. Water cover allowed to accumulate to 3 mm (0.125 in.); and
  3. Runoff simulation allowed to pond to a 25 mm (1 in.) head.

[2]Inflow rates for each test were:
  Test 1, 0.25 gpm/sq in.;
  Test 2, 1.00 gpm/sq in.;
  Test 3, 4.94 gpm/sq in.
  NOTE: gpm × 5.45 = m³/d.

quantifying I/I sources as well as knowledge of the sewer involved in the study.

EVALUATION OF INFILTRATION

Infiltration quantification techniques

Total system. There are various infiltration quantification techniques that can be used to estimate the amount of infiltration entering a sewer system. Because the initial interest with infiltration quantification is with the entire sewer system, these techniques are more applicable to the entire sewer system than they are to a particular sub-system. The following techniques can be used to estimate the total infiltration in a sewer system. The availability of information for the system will have a significant bearing on the method used to estimate infiltration. Some of these methods will use data that, in many instances, will include inflow as well as infiltration. If inflow is a significant portion of the total extraneous water gaining access to the sewer system, additional estimates will need to be made for inflow, and these quantities subtracted from the totals obtained by use of the methods described here. In many systems, where infiltration is the prime source of extraneous water, however, inflow generally makes up only a very small fraction of the total volume of extraneous water gaining access to the system.

Water use evaluation. The water use evaluation method makes use of water supply records for the purpose of estimating the amount of domestic wastewater discharged to the sanitary sewer system. This requires obtaining monthly water-use records for the area being studied. Then, an estimate of the portion of the water that is reaching the sanitary sewer could range from 70% in summer months to 90% in winter months. This then gives an estimate of the domestic, industrial, and commercial wastewater flow rates. With this information, the flow rates can be subtracted from the total flow measured at the area's wastewater treatment facility to obtain an estimate of the infiltration entering the sewer system.

The amount of water that eventually will reach the sewer system is dependent upon the outside-of-house-water use being considered. The amount of water used for lawn and garden irrigation must be estimated. It is possible that some of the water applied to lawns and gardens could gain access to the sanitary sewer system; however, this is difficult to estimate in most instances, and is neglected.

Another factor that needs to be considered when working with water records, is the amount of unaccounted-for water. The unaccounted-for water is generally considered to be the difference between the total water supplied to the water system through wells, springs, or reservoirs and the amount of water measured from each of the individual water meters on each user's water connection. Although water

EVALUATION/REHABILITATION

distribution system leakage can be a major reason for unaccounted-for water, there are other major causes. Incorrect or inaccurate meter reading can be a major cause of unaccounted-for water. This can be a result of inaccuracies of either the meters measuring the input to the system or the individual user's meter. In the latter case, water that is not properly measured could still be directly entering the sewer system, whereas, in the case of leakage, water would not be entering the sewers directly. Illegal taps and unmetered withdrawals from fire fighting lines, street flushing fire lines, and hydrants are other sources of unaccounted-for water. Water from illegal taps could easily enter the sewers, although water from unmetered use of fire lines and hydrants may or may not be entering the sewers.

All of these factors will have a bearing upon the estimate of the amount of water that eventually becomes wastewater. These factors need to be considered in each individual case before estimates are made. If an area is supplied with a secondary water system, thus eliminating the need to use water for outside watering, the amount of water reaching the sanitary sewer system will be a large fraction of the total water supplied to the system. This fraction could be in the 85 to 90% range.

BOD evaluation. This estimating method makes use of the mass BOD loading to estimate the domestic and industrial flow contributions. Once these flow contributions are determined, they are subtracted from the total flow measured at the treatment facility, to obtain an estimate of the infiltration flow rate. Even though the BOD parameter is directly related to the number of people contributing to the sewer system, use of a reactive substance for estimating purposes has been questioned. Experience has shown, however, that the mass loading of BOD is a relatively accurate criteria for estimating domestic flow contributions on a system-wide basis. This method, however, could have severe limitations when applied to small sub-systems.

The monthly treatment plant operating data are used as input information for this method. The average monthly influent flow rate in m^3/d (mgd) and the BOD concentration in mg/L are used. The typical BOD concentration from a domestic wastewater source is 200 mg/L. Typically, this method requires an estimate in the area being considered. An estimate of the wastewater flow from industrial sources also is required.

The following is an example of the procedures used to obtain infiltration estimates with the BOD evaluation method.

EXAMPLE:

Average influent BOD	101 mg/L
Average influent flow rate	103 300 m^3/d (27.3 mgd)
Estimated industrial flow rate	8 700 m^3/d (2.3 mgd)
Estimated industrial BOD loading	4 500 kg/d (10 000 lb/day)

Total BOD loading $= 101$ mg/L $\times$ 103 300 $m^3/d \times 10^{-3}$
$= 10\ 430$ kg/d (23 000 lb/day)

Domestic flow rate $= \dfrac{(10\ 430\ \text{kg/d} - 4\ 500\ \text{kg/day})\ 10^3}{\times\ 200\ \text{mg/L}}$
$= 29\ 500\ m^3 d$ (7.8 mgd)

Estimated infiltration $= 103\ 300\ (8\ 700 + 29\ 500)$
$= 65\ 100\ m^3/d$ (17.2 mgd)

This procedure can be carried out for each month of a year to obtain the total infiltration for the year.

Maximum-minimum daily flow comparison. The theory behind this method is the principle that infiltration is constant throughout any given day. If there is no precipitation, then the daily flow increase is strictly attributable to the domestic flow contribution. Industrial flows are also assumed to be constant throughout the day. If this principle is applied to an entire month, infiltration can be assumed to be constant and the variation between the maximum and minimum daily flows is assumed to be the domestic flow rate.

The above principle can be applied to treatment plant influent flow data to estimate the amount of domestic flow. To determine the estimate for infiltration, the domestic flow is subtracted from the total flow at the treatment plant and the industrial flow estimate also is subtracted. The resulting quantity is the infiltration estimate. This procedure can be carried out with monthly averages to obtain the estimated infiltration for the entire year.

Maximum daily flow comparison. This method is based on the principle that if infiltration does exist in the sanitary sewer system, the influent flow rate measured at the treatment plant cannot drop any lower than the rate of infiltration. This suggests that if the entire population of the area being considered was sleeping, all the industries were shut down, no commercial functions were active, and no precipitation was present, the flow to the treatment plant would be entirely infiltration, because there would be no domestic, industrial, or commercial contributions. For the sake of simplicity, this method can be assumed to occur during the early morning hours in the service area. That infiltration remains constant throughout any given month or week, depending on the available data, also should be assumed.

A graphic plot should be made of the minimum daily flow experienced at the treatment plant for each month during the period of interest. It should be kept in mind that the quantity plotted should not be the average minimum flow for the month, but the lowest minimum daily flow seen during that month. If these points are plotted for each week or month during the year, the difference between the maximum and minimum points on this curve can be assumed to be the annual infiltration for that year. The maximum point on the curve can be considered to be the high groundwater infiltration rate and the minimum point the low groundwater infiltration rate. A good check would be to verify that low infiltration does coincide with dry periods.

The basic drawback of this method is that it does not properly consider the low groundwater infiltration. The low groundwater infiltration could represent a significant portion of the total annual infiltration. Therefore, the closer the low groundwater infiltration rate is to zero, the more accurate this estimating method will be.

Night time domestic flow evaluation. This estimating procedure is based upon a method of flow prediction[5] outlining the relationships of maximum and minimum daily flows with average daily flows based on population. These relationships are for domestic wastewater flow only and are considered to be a fairly reliable tool for estimating.

Figure 2.5 represents a typical dry weather day and shows how the various components of the total wastewater flow are related. The assumptions made for this graph are that the infiltration, industrial, and nighttime domestic flows (NDF) remain constant throughout any given day. The NDF is defined as that portion of the total minimum daily flow that is attributed to domestic activities.

Small sub-systems. The best and most accurate method of estimating infiltration from small subsystems is by direct measurement techniques described earlier in this

EVALUATION/REHABILITATION

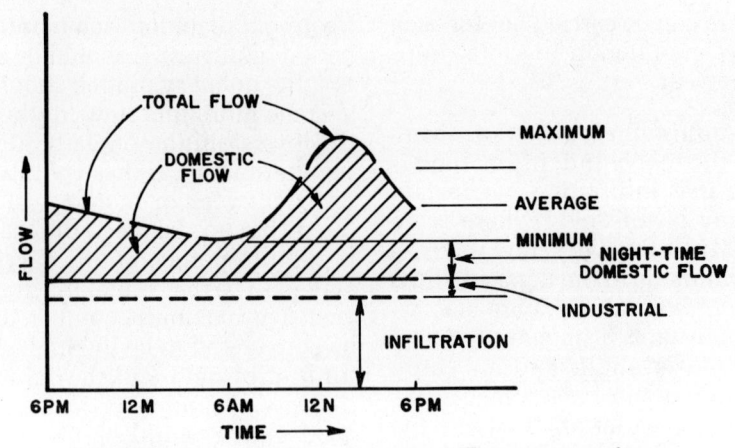

FIGURE 2.5. Daily component flow comparison.

chapter. Some of the estimating methods described here for the total system have severe limitations when applied to small sub-systems.

For instance, water-use data are generally not readily available for small sub-systems. Also, many of the assumptions made with respect to outside watering characteristics become more crucial to the accuracy of the estimates for small sub-systems. Regarding the BOD evaluation method, the quantity of BOD per person per day can vary considerably throughout any city or area of interest. The existence or non-existence of garbage grinders within a small area can affect the BOD loading significantly. These variations can generally be neglected or minimized when considering large or total systems.

The other three methods discussed under total system techniques can be used in small sub-systems; all of them would require direct flow measurements to be taken. If direct flow measurements are taken, however, these methods or variations of these methods can be used to estimate the infiltration from the area of interest.

Once infiltration flow rates or estimates are obtained, unit infiltration or incremental infiltration rates should be calculated. Incremental infiltration is defined as the infiltration flow rate (L/m.d)/m (gpd/in.diam/mile). The calculation of the incremental infiltration requires information regarding the characteristics of the sewer system. The total length of various size pipes within the area of interest must be obtained. Incremental infiltration rates vary widely from city to city and region to region. An incremental infiltration rate for one area may be extremely high, whereas for another area of the country it may be rather low. The significance of the incremental infiltration calculated for any particular sewer system will depend on the characteristics of the area in which the sewer system is located.

Infiltration location. The methods described above help to obtain quantitative estimates of the amount of infiltration in the system. Further flow measurements at key locations within the system can help pinpoint infiltration. Other methods of confirming the results of flow-monitoring techniques are available. These methods can also be used to locate infiltration areas within a system when flow monitoring methods or information is not available.

Temperature. One method that can be used to estimate locations of infiltration is to measure the temperature of the wastewater within the system. This can be done during a manhole inspection, typically a part of a sewer system evaluation. The wastewater can be measured throughout an entire area or sewer system. Temperature data gathered in this manner can provide infiltration information if the temperature of the groundwater can be established. The groundwater temperature can be affected by several factors: the temperature and location of the source of the groundwater, the air temperature and location of the source of the groundwater, the type of material transmitting the groundwater, and the distances the groundwater must travel. The temperature of the groundwater source, in many cases, is a nearby river, stream, creek, or large body of water. Temperatures of these types of waters can be obtained from local 303 (e) studies, 208 Area Wide Water Quality Studies, U.S. Geological Survey Reports and Soil Conservation Service Reports. These documents can also be used to obtain information regarding the soil in the area of the sewer system. Once the temperature of the groundwater is determined, within the area of the sewer system, this can be compared with the sewage temperatures taken during the manhole inspection. This comparison can then be used to project the possible existence of infiltration within an area of interest.

Groundwater levels. This method can be used to confirm flow-monitoring results for a particular area of a system. This method basically helps to confirm flow measurements by relating the groundwater table level to the elevation of the sanitary sewers. Obviously, if the groundwater table is below the sanitary sewers, infiltration will most likely not exist within that area. This may be the case during certain times of the year, because water tables can fluctuate above and below a sewer on a regular cycle.

Interviews. Local contractors who do a considerable amount of underground utility work in the area of interest are a knowledgeable source for groundwater information. These contractors should be consulted to determine areas of severe groundwater problems. Infiltration cannot exist without a continuous water source.

Municipal personnel can also be a valuable source of information on locating general areas of infiltration. Besides being able to provide groundwater information, municipal personnel can provide information relating to backed up sewers and flooded basements within a particular area of a system. Such occurrences can be indications of infiltration problems.

Visual inspection. Visual inspection can be done in different ways. One way would be simply opening manholes within a system during low flow periods, generally at night, and from the surface visually judging depth of flow in the sewer. A more accurate means would be to enter the manhole and measure the depth of flow. Generally, on the upstream portions of a system, this method can be used to eliminate many areas simply by the existence or nonexistence of significant flow during the early morning hours.

Visual inspection can also be accomplished by means of a television camera. This, because of the cost involved, generally becomes the final inspection procedure during a sewer system evaluation. Television inspection, however, particularly when it is done during high groundwater periods, is probably the best way to locate specific points of infiltration. It is difficult, however, to estimate quantities of infiltration by use of a closed-circuit television picture. Direct flow-measuring techniques are still the best way to accomplish estimates for quantities of infiltration.

Cost effectiveness. By simply measuring infiltration and computing the incremental infiltration rates, cost-effectiveness cannot be determined. EPA has attempted to define various incremental infiltration rates that indicate whether removal of infiltration is cost effective. These estimates can be used as a general rule of thumb; however, the only way to determine the cost effectiveness of infiltration removal is by conducting an analysis for each particular sewer system. This becomes a very individual evaluation for each case and is the only way to accurately determine cost effectiveness.

EVALUATION/REHABILITATION

EVALUATION OF INFLOW AND RAINFALL-INDUCED I/I

Rainfall-induced I/I quantification. Wet weather infiltration is defined as a delayed reaction to rainfall, primarily attributable to wet weather percolation through the soil into defective pipes, joints, connections, or manhole walls. The nature and magnitude of wet weather infiltration, like inflow, is heavily dependent on local conditions, including soil types, rainfall characteristics, and the depth and conditions of the existing sewer system.

Because the reaction to rainfall is so dependent on local conditions, comparing rainfall reactions from one system to another is virtually impossible. In fact, collection systems may react differently to similar rainfall events, depending on the preceding weather conditions. As an example, the wastewater flow reaction to a specific rainfall would probably differ greatly if the rainfall was preceeded by several weeks of dry weather as opposed to several weeks of wet weather. Therefore, a direct correlation of measured rainfall to rainfall-induced I/I entering a particular system is almost impossible without many years of historical data.

Another important factor in I/I quantification is surcharging. Regardless of the amount or intensity of rainfall, once a system surcharges, the maximum peak inflow rate has been achieved for that particular storm. The peak inflow rate is ultimately controlled by the size and head of the leaks. Some systems surcharge very rapidly and require only marginal rainfall events to reach the maximum peak inflow rate, while other systems react more slowly and surcharge only during prolonged, heavy rainfall. This is another indication that each system is unique and must be evaluated separately based on actual hydraulic conditions rather than data from other systems in different locations.

To quantify the extraneous flows entering any collection system accurately, the total flows must be appropriately measured or at least accounted for and estimated, including contained flows (remaining in the system) and escaping flows (such as overflowing manholes, bypasses). Complete and accurate flow monitoring is extremely important if I/I quantities are to be determined. The contained flows should be measured using one of the measuring techniques described in Chapter 3. The escaping flows also should be measured or at least estimated. If hydraulic conditions make measurement impossible, then the location, type, and duration of the escaping flows should be noted and an estimate must be included with the measured contained flows.

A common misconception in the quantification of rainfall-induced I/I is that wastewater treatment flow records are the only data required for evaluation. In many facilities, flow measurement equipment is located on the effluent discharge, which would not record bypassed flows at the head of the facility, much less any overflowing manholes or other bypasses within the collection system.

Applicable flow measurement techniques. Many types of flow measurement techniques, described in Chapter 3, are available and the selection of flow measurement equipment should be based on the type of information desired. For example, if flow data from various sections of the collection system are needed, then equipment that provides adequate data using manholes should be selected. Measurement can be made for entire collection systems or for a single leak, depending on the desired results.

A thorough investigation of overflows and bypasses should be discussed with the appropriate employees or former employees to obtain information such as location, duration, type of overflow or bypass, and rainfall events necessary to produce such conditions.

If the location of the bypasses (installed pipes to relieve surcharging or overflowing) is known prior to the contained flow measurement period, the pipes can be plugged, if feasible, to allow for complete flow measuring. If plugging creates adverse conditions (such as wastewater backups into residences) then the bypassing must be estimated or measured so that a bypass rate and quantity can be included in the final flow data.

Devices that adequately measure rainfall range from sophisticated electronic equipment to simple manually-read rain gauges. The rainfall data can be presented hourly, daily, or at whatever time interval is appropriate, depending on the equipment. Regardless of the type of equipment chosen, rainfall amount, duration, and intensity are important parameters that can be used independently or collectively to interpret more accurately rainfall-induced I/I quantification. Provided there are staff in the vicinity, the most economical rainfall measurement technique is simply reading a rain gauge hourly or daily. The gauge should be located in that area of the collection system under consideration for an accurate comparison of rainfall data to wastewater flows.

Flow measurement methodology. If techniques can be defined as tools, then methodology can be defined as how those tools are used to compare, analyze, and evaluate the collected data and other pertinent information in order to formulate results and conclusions. The wastewater flow data, measured rainfall, and existing conditions represent the tools. The type of wastewater flow data and rainfall data desired should depend on the financial resources available and to what extent the data will be used. A simple inexpensive study can relate daily rainfall to daily flows to determine if a problem does exist and to what magnitude. More sophisticated and elaborate equipment will generate more extensive data, but also will cost more. Each community should implement a program that will produce the minimum needed results to generate rehabilitation measures for the least cost.

The raw wastewater flow data can be converted into several very useful flow parameters, including average dry weather flow (Q_D) dry weather peak flow (Q_{DP}) wet weather maximum day flow (Q_{WM}) and wet weather peak flow (Q_{WP}). Other parameters can be developed on a case-by-case basis. The average dry weather flow can be defined as the average flow that occurred on a daily basis that had no evident reaction to rainfall. The dry weather peak flow can be defined as the highest measured hourly flow that occurred on a dry weather day. The wet weather maximum daily flow can be defined as the maximum flow measured over a 24-hr period. And finally, the wet weather peak flow can be defined as the highest hourly flow measured during wet weather. The wet weather flows must include measured or estimated overflows and bypassed quantities.

Flow data can also be used to project annual flows if the study period is less than a year in duration. The most common presentation method of the flow data is with the use of hydrographs, which can include daily flows, hourly flows, and rainfall data. Each hydrograph can be developed uniquely to show any specific type of data required. A typical hydrograph is shown in Figure 2.6. Note that the hydrograph includes escaping flows. The immediate increase in daily flows followed by an immediate return to the dry weather flow range indicates inflow, whereas increased flows that remain elevated for several days usually indicates wet weather infiltration.

These basic flow parameters have several important uses. The average dry weather flow can be used to size the average design capacity of a wastewater treatment facility. The wet weather maximum daily flow can be used to size equalization facilities prior to treatment. The wet weather peak flow rate is important in sizing collection lines and lift stations, and providing adequate hydraulic capacity for individual treatment units.

The basic flow parameters can also be used to quantify the rainfall-induced I/I. The difference between the wet weather maximum flow and the average dry weather flow can obviously be defined as the maximum daily I/I. The difference between the wet weather peak flow rate and dry weather peak flow rate would be a conservative rainfall-induced peak rate.

EVALUATION/REHABILITATION

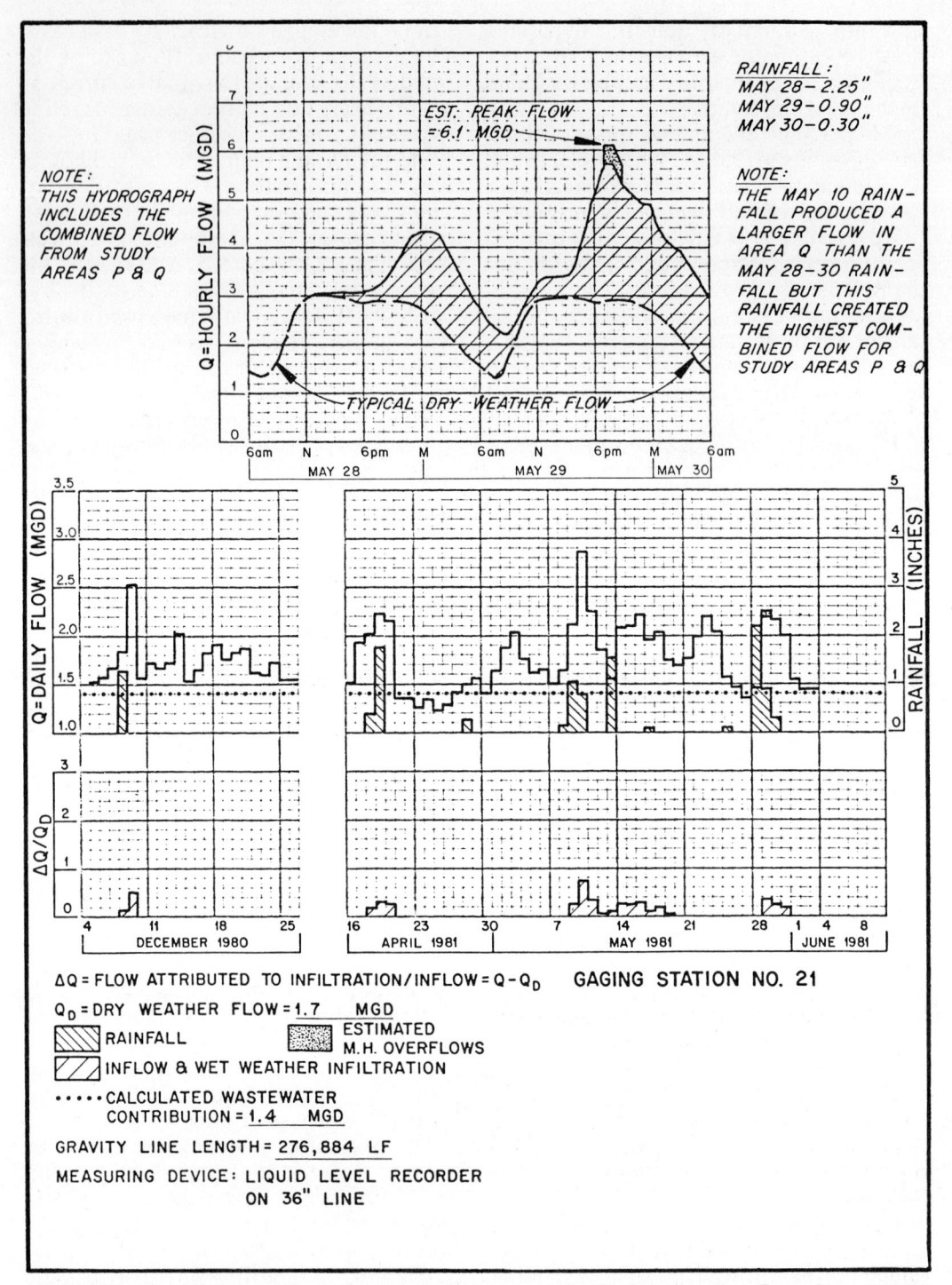

FIGURE 2.6. Hydrographs—the most common method of presenting flow data. NOTE: gpm × 5.451 = m³/d; in. × 25.4 = mm.

The wet weather flow also can be compared to average dry weather flow to develop ratios that help to define the magnitude of the I/I problems during wet weather. Even in well constructed separate systems, the ratio of wet maximum to dry average (Q_{WM}/Q_D) usually ranges from 2 to 3, and the wet peak to dry average (Q_{WP}/Q_D) from 3 to 4. Higher values obviously indicate a more pronounced problem. These ratios only define magnitude, they do not affect cost-effectiveness. In some cases, the ratios may be 2.0 to 3.0, respectively and some rehabilitation may be required; in other systems, the ratios may be higher and rehabilitation may not be cost-effective. The economics of rehabilitation is not based on flow, but solely on cost: cost of transportation, cost of treatment, and cost of operation and maintenance.

Another important aspect of rainfall-induced I/I is the correlation to various types of rainfall events and durations. Each collection system is unique in operation. In other words, no two systems operate identically because soil type, topography, area rainfall characteristics, and collection system condition can vary significantly. Therefore, it is impossible to compare the wet weather flow reaction of one system to another. And, it is also impossible to predict a reaction for one system to a particular rainfall event based on existing data from another system using a similar rainfall event.

Even within the same system, identical rainfall events may produce totally different wastewater flow reactions. A certain rainfall episode could be preceeded by several weeks of extended wet weather flow as opposed to several weeks of dry weather. In one case the ground is very dry and capable of absorbing the rainfall into the soil matrix. In the other case, the soil is probably saturated and would have an entirely different effect on the wastewater flows.

It is difficult to predict with any degree of confidence a flow reaction from a larger rainfall event based on data from a smaller event. In other words, wastewater flows and rainfall intensity do not have a linear relationship. In all cases, there is a point at which all inflow sources are activated and a maximum inflow rate essentially is reached, regardless of the amount of rainfall. Extended rainfall can only increase the total quantity entering the system, which would be reflected in an increase in annual flows, but not the peak flow.

In summary, the quantification of rainfall-induced I/I for entire collection systems or segments of collection systems can only be accomplished with actually measured flow data including contained and escaping flows. The flow data must be correlated with measured rainfall and the evaluation process must be limited to the collection system in question and for only the range of the rainfall data measured.

Rainfall-induced I/I localization. Rainfall-induced I/I enters wastewater collection systems from three primary sources: 1) collection lines, 2) manholes, and 3) service lines. Several different types of leaks can be associated with each primary source. These type of leaks include:

1. Collection lines
   - Point source leak;
   - Leak at a defective joint;
   - Point source leak at service tap;
   - Multiple leaks along entire line;
   - Storm drain cross-connection.

2. Manholes
   - Through holes in cover and between lid and ring;
   - Under ring only;
   - Within top two feet of manhole wall;
   - At perimeter of manhole floor only;
   - Throughout manhole wall;
   - At broken or missing manhole cover;
   - At broken or missing manhole ring;
   - At stub-out in manhole;
   - Around pipes.

3. Service lines
   - Point source leak;
   - Multiple leaks;
   - Leak at missing or defective cleanout

EVALUATION/REHABILITATION

or plug;
- Leak on discontinued line;
- Storm drain cross connection;
- Foundation drains;
- Connected foundation drain;
- Roof drain connection;
- Surface drain connection.

Other types of leaks also may exist on the three primary sources. Also, the types of leaks will vary widely depending on local conditions and the area of the country. One part of the country may experience rainfall-induced I/I primarily from drains connected to service lines, whereas another part of the country may experience inflow from defective collection lines located in drainage ditches or rapidly draining soils. Rainfall-induced I/I sources generally will be scattered throughout every collection system and cannot be isolated in small portions of the system with any degree of accuracy. Eliminating large portions of a collection system without smoke testing the area will significantly reduce the overall effectiveness of the sewer system evaluation.

Applicable techniques. Several techniques currently exist for locating rainfall-induced leaks including 1) smoke-testing, 2) dye water testing, and 3) leak quantification. These techniques are discussed in an earlier section of this chapter.

Methodology. The applicable techniques of rainfall-induced I/I localization can be used independently or in conjunction with one another. Smoke testing will locate leaks generally by collection line, manhole, or service line. Dye-water testing will more specifically locate the type of leak, especially on manholes. The internal inspection technique will locate collection leaks and define the type, if used in conjunction with flooding.

Cost effectiveness. Regardless of the amount of extraneous stormwater that enters a sewer system or the quantity that can be identified for elimination (usually expressed as a percentage of the total I/I), the ultimate factor that defines excessive rainfall-induced I/I is cost. The measured wastewater flows including I/I only establish the following cost parameters: 1) survey and rehabilitation costs, 2) construction and capital costs, and 3) operation and maintenance costs. These cost parameters are used in the economical analysis, which will establish what percentage, if any, of the total rainfall-induced I/I is cost-effective to eliminate (Table 2.2).

The costs associated with survey and rehabilitation can be defined as survey tasks and types of rehabilitation. Rainfall-induced I/I will also affect the need for lift stations, relief lines, and treatment units. The construction and capital costs of these items must be incorporated in the cost evaluation process. The operation and maintenance costs also will be affected by the extraneous wet weather flows, pumping costs, energy usage at treatment facilities, and chemical usage.

A comparison of the cost to transport and treat the rainfall-induced infiltration/inflow with the cost to eliminate it, should yield the most cost-effective solution.

The basic survey cost for the evaluation of rainfall-induced I/I ranges from $1.15 to $1.65/m ($0.35 to $0.50/linear ft, 1st Quarter, 1983). This cost includes money for smoke testing, dye-flooding, visual inspections, limited cleaning and TV inspection, and recommendation for rehabilitation.

Figure 2.7 is a typical plot of transportation and treatment cost curve, which includes the operation and maintenance costs and the cost of construction of all facilities that are affected by I/I. This curve obviously shows a reduction in overall cost as the amount of I/I is reduced.

Figure 2.7 also shows the survey and rehabilitation cost. As shown, the cost of repair increases significantly if all leaks were potentially eliminated.

By combining these two curves, a composite cost curve can be developed. This curve has a minimum cost point that indicates the maximum amount of I/I that can be cost-effectively eliminated.

TABLE 2.2. Cost of rehabilitation for elimination of inflow.

Project	Sewer Length		Leaks Repaired[1]		Inflow Removed[2]		Bid Opening Date	Bid Price ($)	Contractor Cost of Repair		
	(L.F.)	(miles)	(total)	(per mile)	(gpm)	(gpd/mile)			($/leak)	($/gpm)	($/L.F.)
Bastrop, LA	492 730	93.3	486	5.21	5053	77 970	10/13/81	345 832	712	68.44	0.70
Center, TX	181 880	34.4	181	5.25	865	36 160	3/18/81	40 440	223	46.75	0.22
DeRidder, LA	270 220	51.2	249	4.87	2992	84 180	10/25/82	254 717	1023	85.13	0.94
Farmersville, LA	114 700	21.7	161	7.41	902	59 790	6/14/82	81 816	508	90.71	0.71
Grand Saline, TX	90 500	17.1	104	6.07	1110	93 250	9/30/81	92 914	893	84.00	1.03
Mandeville, LA	182 480	34.6	274	7.93	932	38 830	12/4/80	187 233	683	200.89	1.02
Sulphur, LA	489 160	92.6	931	10.05	5562	86 450	11/13/79	619 276	665	111.34	1.27

[1]Majority of the leaks repaired were inflow sources detected through smoke testing and considered cost effective to repair. Does not include repair of leaks on private property. The type of repairs varied from one system to the other but overall they fell into the following categories:

1. Point repair of sewer and service lines
2. Replacement of a section of pipe or entire manhole reach
3. Installation of parallel lines
4. Sliplining
5. Pressure testing and grouting of defective joints
6. Replacement of pipe section to eliminate cross connection
7. Sealing of manhole wall
8. Sealing at junction of frame and MH wall
9. Sealing at the junction of MH wall and base
10. Rebuilding of MH wall
11. Replacement of MH cover and/or frame
12. Replacement of cleanout risers
13. Plugging of stubouts
14. Sealing around incoming lines at manholes
15. Plugging of abandoned service lines
16. Replacement of a section or entire service line
17. Disconnection of roof drain/surface drain
18. Repair or replacement of wyes, tees or pipe/pipe connections

NOTE: mile × 1.609 = km; gpm × 6.308 × 10^{-2} = L/s; gpd/mile × 2.352 = mL/m·d; $/gpm × 15.85 = $/L·$s^{-1}$; $/L.F. × 3.281 = $/m.

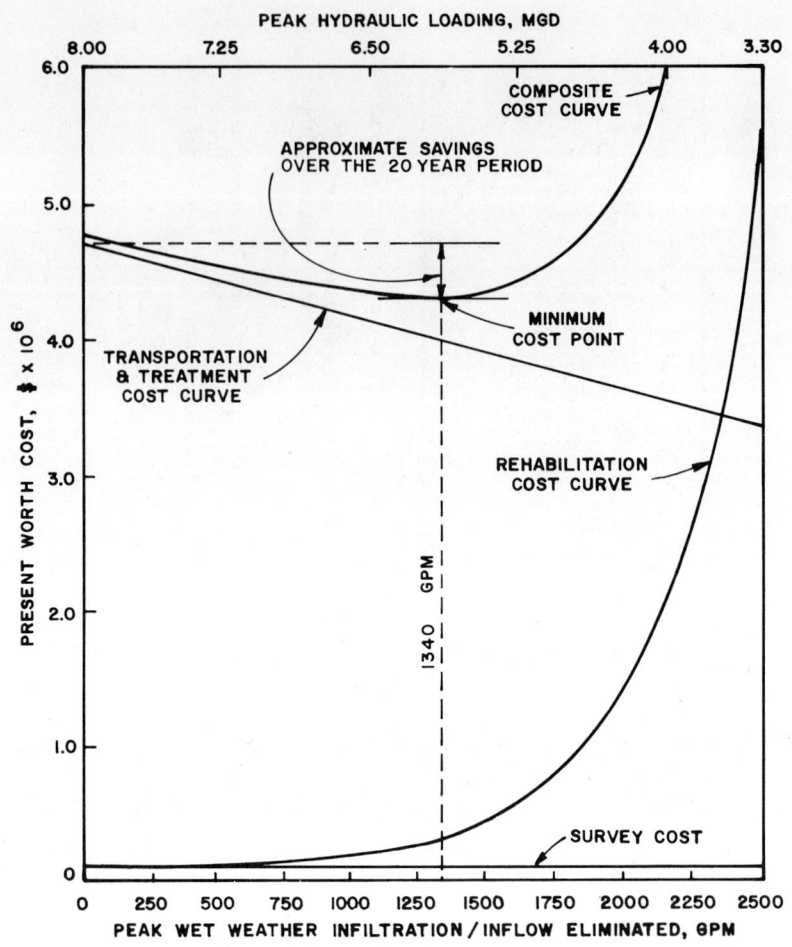

FIGURE 2.7. Cost curve for determination of I/I sources to be rehabilitated. NOTE: gpm × 5.451 = m³/d; mgd × 3785 = m³/d.

EVALUATION OF PHYSICAL CONDITIONS

Structural integrity. When any evaluation is undertaken on a sewer collection system, an analysis for structural integrity and defects should be included. In many areas the conditions of overburden may have changed over the years because of changes in land use. In some instances, the existing overburden may be more than the original pipe or manhole wall was designed to accept. In addition, some changes in areas may have resulted in traffic loads that were not considered in the original design of the sewer. Because there is a wide range of rehabilitation alternatives, pipe strength and manhole stability must be considered so the best alternative of rehabilitation can be selected. The depth limitations on pipes and manholes vary according to the type of material and its related structural properties.

The second area of structural integrity involves the existing condition of the pipe itself. The conditions of the pipe can have an impact on both hydraulic capacity and bearing strength of the pipeline. Each type

EVALUATION METHODS

TABLE 2.3. Information gathered when examining for structural defects.

Pipelines	Manholes
Type of pipe	Type of material
Horizontal deflections	Condition of walls
Vertical deflections	Condition of joint between manhole and frame
Joint separation	
Broken joints	Condition of channel
Root penetration in joints	Condition of joints of pipe connections
Circumferential cracks	
Longitudinal cracks	Condition of joints on risers
Crushed pipe	Any visible leakage
Missing pipe	Corrosion of walls
Protruding taps	Condition of frame and cover
Broken taps	Condition of steps
Leaking joints	
Leaking taps	
Holes in pipe	
Corrosion of pipe walls	

of defect must be evaluated to determine its related impact on strength and capacity so the proper rehabilitation method can be chosen. When inspecting for structural defects, the information shown in Table 2.3 should be gathered for evaluation purposes.

Operation and maintenance related problems. Every utility experiences some operational and maintenance problems that can be attributed to specific causes. The problems that are indicators of sewer pipe or manhole defects are listed below:

- Overflowing manholes
- Sewer backup in buildings
- Surcharged sewer
- Exfiltration of wastewater
- Sunken areas above sewer line
- Increased flows
- Sand and gravel in wetwells
- Stuck flap gates or open overflows

Detection of any of the above indicators should lead to further inspection to determine the cause of the problem. The problems may be called in from the public or would be detected by routine inspection programs of the utility. It is imperative that all information reported is recorded by location so the problem history for a sewer line or manhole can be correlated with the specific manhole or sewer reach. Extensive training of employees is necessary to ensure that they collect the necessary information needed for rehabilitation decisions. Most operation and maintenance employees concentrate on correction of the immediate problem without visualizing the total permanent solution of some type of rehabilitation.

Map correction and updating. The backbone of any good sewer collection system are the permanent records that indicate the skeleton of the piping and the location of manholes in the system. Each utility should have detailed plans of the specific construction contracts so that exact location of the sewer lines and manholes can be determined. All drawings should be to a specific scale so that scaled measurement to manholes is possible from other structures in the area. In addition, there should be smaller scale drawings of the entire sewer collection system so that general patterns of the collection system are evident.

A specific procedure must be established by the utility for correction of errors and updating the drawings. The field personnel must be properly oriented to recognize discrepancies in the field conditions and the maps. They must also be properly trained to record any changes that are necessary to correct the existing mapping system. All corrections and changes should be coordinated by the centralized records section of the utility. All corrections and changes made to the records should be accomplished in a timely manner and the new drawings distributed to the field personnel. The accuracy of drawings used by the field personnel is critical to proper identification of sewer collection system components. Without proper mapping the sewer collection system becomes unduly difficult. The strict procedures for map formulation and updating are very cumbersome and time-consuming for the utility, but are critical for proper operation and maintenance of the sewer collection system.

The mapping system must have proper identification of the sewer lines so that the detailed construction plans can be located quickly. In addition, there should be a specific method of identifying each manhole. By having all manholes individually identified, each sewer reach can then be individually identified. The identification of each manhole and sewer reach is imperative if the maintenance history and physical conditions are to be correlated properly. A method of properly identifying relocation work and rehabilitation work on the sewer collection maps must be developed. By proper identification of all manholes and sewer reaches, a sewer system inventory can be developed.

Applicable techniques

Manhole and pipeline visual inspections. Visual inspection of manholes and pipelines can involve both a surface inspection program and an internal inspection program. A good number of problems can be detected by a visual inspection when walking along the sewer right-of-way. Particular attention should be given to sunken areas over the sewer, areas with ponding water, condition at stream crossings, condition of manhole frame and cover, condition of any exposed brickwork, and whether the manholes or special structures are visible. As stated above, the importance of individual identifications for each structure is extremely important for correlating the conditions during the surface inspection.

The internal inspection program involves entering the manholes and visually reviewing the condition of manholes, channels, and pipelines. Each manhole should be visually inspected for the following conditions:

- Seal of frame to manhole shaft;
- Condition of manhole shaft between frame and manhole cover;
- Structural condition of walls;
- Joint condition in precast manholes; and
- Any visual leakage.

Each bench and channel portion of a manhole should be inspected for the following conditions:

- Condition of manhole shaft;
- Any leakage in channel;
- Any leakage between manhole wall and channel;
- Any damage or leakage where pipeline connects to manhole; and
- Any flow obstructions.

Some pipelines can be inspected with lights or mirrors. Visual inspection in smaller sewers is limited in the scope of problems detected. Generally, the only portions of the sewer that can be seen in detail are those close to the manholes. The remainder of the line can be inspected by mirrors to determine any horizontal or vertical alignment problems and any large leaks. The visual method of inspection on smaller sewers will not give the inspector definitive information on cracks or other structural problems. Even though there are drawbacks to visual inspection of smaller sewers, it allows the gathering of information necessary in making rehabilitation decisions. Larger sewers can be visually inspected by walking.

Television inspections. Closed-circuit television equipment has become the most effective method of inspecting the internal condition of sewers. Not only can reports be generated with the inspection, but a permanent visual record can be made for subsequent review. The range of inspection is unlimited; smaller cameras will even televise 4-in. service laterals. When using television equipment, the operator must be properly instructed so that required data can be obtained. When recording the data on report forms, the operator must be able to interpret the visual picture and translate this information to a physical work description. It is imperative that closed circuit television operators review the conditions with equal perspective so that results from different operators can have the same basis for comparison.

The technology has evolved to the point that both black and white and color closed-circuit television units are available. The one advantage of color is additional depth perception. The black and white pictures give plenty of information and are used extensively for detection of physical conditions in sewer lines.

It is necessary to set up a routine recording procedure so that the television inspection reports can be easily referenced to the particular sewer reaches. In addition, if the reach of sewer is videotaped a cross file must be established for future review. Some items that may be recorded on a television inspection report are listed below:

Television inspection data
- Length of section;
- Type of pipe;
- Joint spacing;
- Root penetration (location, quantity rating);
- Grease sediment presence (location);
- Horizontal deviations (location, length);
- Open joints (location, severity);
- House connections (number, location, condition);
- Water levels;
- Circumferential cracks (location, severity);
- Longitudinal cracks (location, severity); and
- Missing pipe (length, severity).

REFERENCES

1. Nogaj, R.J., and Hollenbeck, A.J., "One Technique for Estimating Inflow with Surcharge Conditions." J. Water Pollut. Control Fed., **53**, 4, 491 (Apr 1981).
2. "Manual on Uniform Traffic Control Devices for Streets and Highways." U.S. Dept. of Transportation, U.S. Gov't. Printing Office, Washington, D.C.
3. "Operation and Maintenance of Wastewater Collection Systems." Water Poll. Control Fed., Manual of Practice No. 7, Washington, D.C. (1982).
4. "A Report on Inflow of Surface Water Through Manhole Covers." Nennah Foundry Co., Neenah, WI.
5. "Design and Construction of Sanitary and Storm Sewers." Water Poll. Control Fed., Manual of Practice No. 9, Washington, D.C.; Amer. Soc. Civil Engr., Manual of Engineering Practice No. 37, New York, N.Y. (1974).

Chapter 3

FLOW MONITORING

43	General Data Needs
	   Study Area Maps
	   Existing Sewer System Records
		56	Design of a System Monitoring
			   Program
			   Research Existing Information

45	Flow Measurement Techniques
	   Manual Methods
	   Automatic Flow Meters
	   Velocity
	   Groundwater Measurement
	   Rainfall Measurement
		62	*Site Selection*
			   Safety Program
			   Monitoring Necessities
			   Monitoring Duration
			References

GENERAL DATA NEEDS

Study area maps. It is essential to have maps of the study area before starting any system study. Some useful information, and in many cases, extremely valuable information, may be obtained from USGS topographic maps; state organizations such as a department of natural resources, geologic survey, and health department; regional planning organizations; county governments, sewer districts, and utilities companies.

A review of these data before or very early in the study program will serve as a valuable orientation to the area. Also, potential trouble areas such as altered drainage paths, filled lakes and streams, and areas of high groundwater may be identified.

Existing sewer system records. Every organization that has responsibility for the sewer system has some type of record-keeping system. These record systems run the complete gamut from handwritten notes on scrap paper to very sophisticated computer data retrieval systems. Regardless of how the records are kept, it is essential that they be collected, sorted, reviewed, and utilized to provide a meaningful and effective program.

Sewer maps. If the sewer system study is to be conducted properly, a complete, up-to-date set of sewer maps is essential. Ideally, these maps should be to a scale that is convenient for both office and field use.

Sewer condition. Information about sewer condition includes such items as: date of construction, methods of construction, type of pipe, type of manhole, type of joint material, sewer size, sewer depth, collapsed sewer areas, and chemical damage areas. This information should be supplemented by information about complaints from the residents (type and frequency of problem) and meetings with maintenance personnel. In any event, the maintenance personnel should not be overlooked. In many instances they may be able to define problems and answer many questions because of their working knowledge of the system.

Collection system flow records. Any actual sewer flow information that is available should be collected, verified, and analyzed. In some instances, there may be very little information available. Sometimes, daily (or less frequent) manual flow readings at the treatment plant may be the only flow information that can be collected. On

EVALUATION/REHABILITATION

the other hand, a wealth of information may be found in years of continuous flow data at the treatment plant.

Occasionally, subsystem flows may be determined from pump station records and from instantaneous or short duration flow records from previously conducted metering programs. If large industrial or commercial users are located within the study area, they may have sewer discharge records available.

Bypass and overflow information. Almost all sewer systems have some type of bypass or overflow condition. Bypasses and overflows are usually constructed to relieve a sewer flooding or back-up condition. Sometimes, however, they act as a natural system relief where raw wastewater is discharged from a manhole cover because of severe system surcharging.

Both bypasses and overflows must be documented as to location and frequency of discharge. During some portion of the system study, quantification of their discharge is essential in order to accurately assess the sewer flow conditions.

Emergency pumping. In many sewer systems during storms or high water conditions, emergency pumps may operate to prevent sewer back-up and overflows. Quantification of the pumping activity gives the analyst a better understanding of the functioning of the collection system. During a flow-monitoring program, the time, duration, and quantity of emergency pumping must be determined and taken into consideration to establish true system flows.

On occasion, construction within the study area can have associated pumping or dewatering activities. These activities can affect sewer flows by reducing groundwater levels. Also, in many instances during these types of activities, discharge is pumped directly into a sewer manhole, legally or illegally, and drastically distort the flow in that particular area.

Water usage. The quantity of water used by residential, commercial, and industrial users is usually documented by water meter readings. These usage records are very helpful in determining the expected wastewater flow in the collection system (baseflow). Although not always available and often distorted by private water supplies (wells and stream intakes), that information is necessary for complete understanding of the collection system.

Rainfall data. Rainfall data for the area covered by the collection system can be useful for determining relationships between sewer flow events. Rainfall can also help explain some of the unusual flow situations that are always present in any system study. Amounts of rainfall and the variations across the study area are an essential element of any flow-monitoring program.

The type of rainfall records available may vary from an official weather station to an event gauge at the treatment plant. It is often necessary to establish a rain gauge network as a part of the sewer system study.

Groundwater, lake, and stream data. Groundwater, lake, and stream elevations can play an important role in infiltration and inflow quantities. Consequently, any record of these elevations is important to the understanding of the sewer system operation. These records may be collected by a variety of organizations including: the U.S. Geological Survey, Corps of Engineers, state agencies, city and county governments and industrial and contracting firms. At times the news media and private individuals also can help, particularly concerning record high water levels.

Demography. Demography or population trends within the collection system study area is another important key to be considered while analyzing records. The past and projected population and population trends may help explain historical flow data and changing flow patterns. These data also should be considered while developing the study plan so that adequate data are gathered to allow proper planning for future population expansion and shifts.

Recordkeeping program. Even under the best of circumstances, the existing records and recordkeeping program will not be able to provide all the needed information. Consequently, there is an opportunity to reassess the current data collection and storage program. In many instances, only minor revisions to the existing program are necessary to provide better and more easily obtained information. On the other hand, a consistent recordkeeping program may need to be initiated at this time. In any case, it is essential that a routine, meaningful, ongoing recordkeeping program be maintained even though a specific study is not in progress.

FLOW MEASUREMENT TECHNIQUES

Monitoring flows within a wastewater collection system is important. The resulting data form the basis for determination of user costs, volume of rainwater or groundwater entering the system, existing line capacity, treatment plant O&M, effectiveness of rehabilitation program, and design of future needs.

Many techniques can be utilized in the measurement of flows in sanitary sewers. The equipment and technique selected will depend upon the resources available, the degree of precision required, and the physical conditions within the sewer. This section presents the more commonly used flow-monitoring methods along with a brief summary describing the technique. Additional information on flow-monitoring techniques and accuracy can be found in the References to this chapter.

Manual methods. Manual methods are the most widely used techniques for measuring instantaneous or short-term flow. Generally, the equipment is portable and flows can be determined immediately using published curves, nomographs, or tables.

Weirs. The weir is a common device for measuring wastewater flows because of its ease of installation and low cost. To design a weir that will provide useful data, the following should be considered:

• The weir should be constructed of a thin plate with straight edge, or thick plate with a knife edge.
• The height of the weir, from the bottom of the channel to the crest should be at least 2 times the expected head above the crest (tends to lower approach velocity).
• All connections between the weir plate and channel must be water tight.
• To prevent a vacuum on the nappe underside, the weir should be ventilated.
• The weir must be level.
• Weir crest and approach channel must be cleaned periodically.
• The head over the weir should be measured at a point located upstream of the weir a distance of at least 2.5 times the head over the weir, at the weir plate, or in a stilling well as long as the point of measurement is not affected by the drawdown of water level approaching the weir crest.
• Weirs should be placed in a straight stretch of sewer with little slope to lower the approach velocity. It may be necessary to utilize baffles for uniform velocity distribution upstream.
• The weir size and type should be determined only after a preliminary field survey has determined the existing and anticipated flow rates.
• Weirs installed under field conditions should be calibrated to ensure accurate measurements.
• Under surcharge conditions or when free fall over the weir does not occur, weir equations and data should not be used.

Measurements are taken by recording the head (water level) above the weir crest and determining flow rates by calculation, nomographs, or tables. Figure 3.1 presents commonly used weirs. Recommended usage for each weir is described below.

Triangular (V-Notch) weir—An accurate device particularly suited to measuring low flows. Best weir profile for discharges of less than 30 L/s (1 cfs) and may be used for flows up to 300 L/s (10 cfs). General operating range when installed in manhole is 0 to 90 L/s (3.0 cfs).

Rectangular (contracted) weir with end contraction—Able to measure much higher

EVALUATION/REHABILITATION

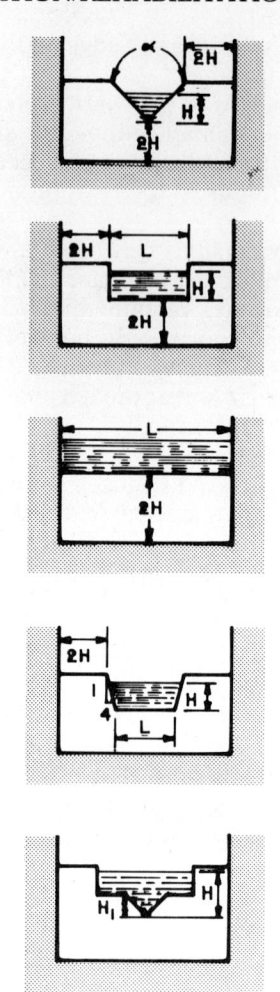

FIGURE 3.1. Flow monitoring weirs. From top: triangular; rectangular contracted; rectangular suppressed; trapezoidal; compound.

flows than V-notch weir. Discharge equation more complicated than those for other types of weirs. Widely used for measuring high flow rates in channels suited to weirs.

Rectangular (suppressed) weir without end contractions—Able to measure same range of flows as contracted rectangular weir, but easier to construct and has simpler discharge equation. Width of weir crest, however, must correspond to width of channel, so use is restricted. May have problems obtaining adequate aeration of nappe.

Trapezoidal (cipolletti) weir—Similar to rectangular contracted weir except that inclined ends result in simplified discharge equation.

Compound weir—Combination of any two types of sizes of above weirs to provide wide range of flows. Ambiguous discharge curve in transition zone between two weirs.

Weirs can be used in conjunction with depth recorders to record head (H) over the weir. Special care must be applied to ensure that the head being recorded is carefully referenced to the weir crest. Also, caution must be used when monitoring flows in sewers. Solids and debris tend to settle upstream of the weir. This deposition can lead to odor and corrosion, may affect the weir length, and may alter the accuracy of the measurements. Figure 3.2. shows an example of rectangular weir construction and installation. Commercially constructed weirs are readily available for most standard sites and locations. Advantages and disadvantages of weirs are listed below.

Advantages—

- Low cost;
- Easy to install;
- Easy to obtain flow by standard equations, nomographs, etc.

Disadvantages—

- Fairly high head loss;
- Must be periodically cleaned; not suitable for channels carrying excessive solids;
- Accuracy affected by excessive approach velocities and debris; and
- May be difficult to make accurate manual measurement in sewers because of limited access.

Flumes. Flumes operate as an open channel form of the Venturi principle. In the flume, the constriction at the throat causes the flow to go through critical depth. This is followed by a hydraulic jump if the slope allows subcritical (low velocity) flow. There are several types of open channel flumes, including the Parshall, Palmer-Bowlus, H-Flume, and Trapezoidal configurations. The hydraulics of the flumes are discussed in the works listed in the

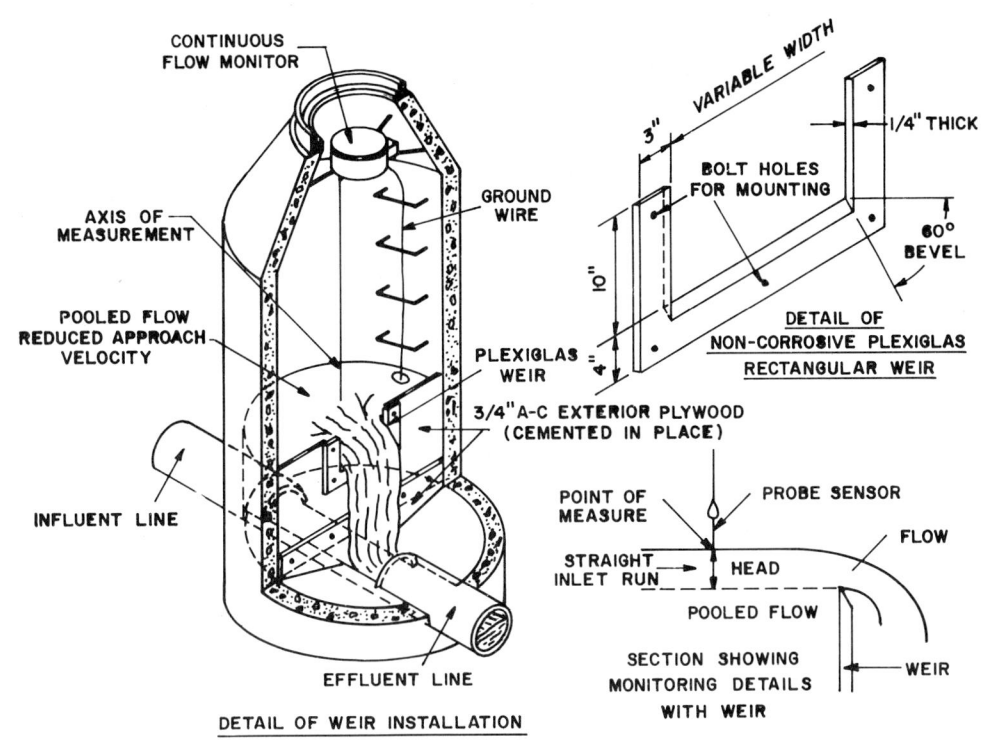

FIGURE 3.2. Weir details and installation. NOTE: in. × 25.4 = mm.

references.[1] Generally speaking, flumes are capable of providing results accurate to within 3 to 5 percent.[2,3] Advantages and disadvantages are listed below.

Advantages—

- Self-cleaning to a certain degree;
- Relatively low head loss;
- Accuracy less affected by approach velocity than it is in weirs;
- Data easily converted to flow using tables or nomographs.

Disadvantages—

- High cost;
- Difficult to install.

Flow moving freely through a flume is calculated from a measurement of upstream water level. Depth-measuring devices can be used to obtain continuous flow data; manual measurements can give instantaneous flow data. Flumes are self-cleaning with no sharp edges to cause setting or clogging, and accuracy is less affected by approach velocity than other measuring devices. These monitoring devices are used widely in measuring influent/effluent at wastewater treatment plants.

General requirements for the installation of flumes are:

- A flume should be located in a straight section of the open channel, without bends immediately upstream.
- The approaching flow should be well-distributed across the channel, and relatively free of turbulence and waves.
- Generally, a site with high velocity of approach should not be selected for a flume installation. If, however, the water surface just upstream is smooth with no surface boils, waves, or high velocity current concentration, accuracy may not be greatly affected by velocity of approach.
- Consideration should be given to the height of the upstream channel, with regard to its ability to sustain the increased depth caused by the flume installation.
- Although less head is lost through

47

EVALUATION/REHABILITATION

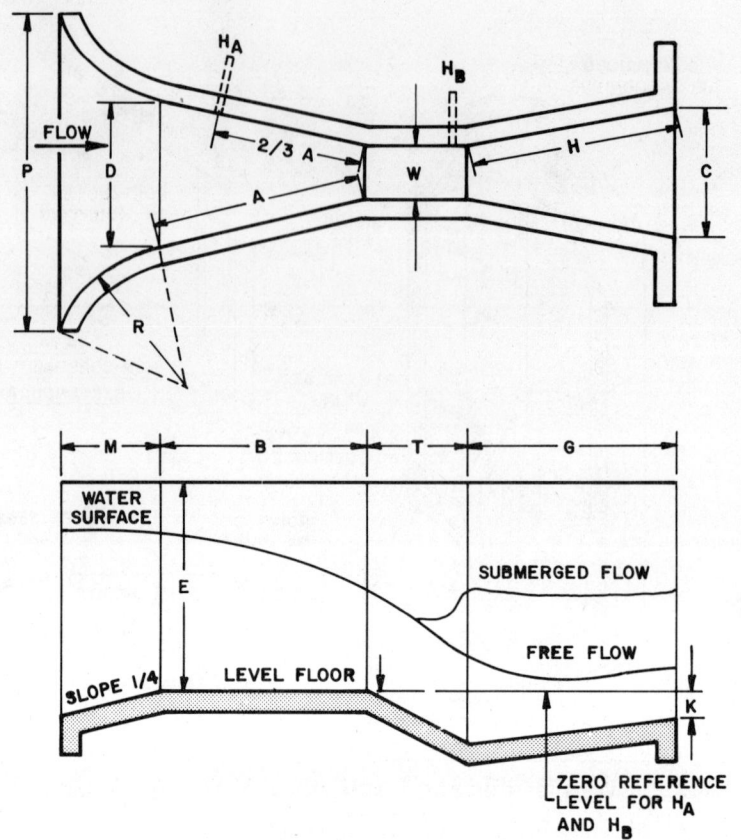

FIGURE 3.3. Parshall flume.

flumes than over weirs, it should be noted that significant losses may occur with large installations.

• The possibility of submergence of the flume due to backwater from downstream should also be considered, although the effect of submergence upon the accuracy of most flumes is much less than is the case with weirs.

Parshall flume—The Parshall flume (Figure 3.3) was originally developed for monitoring irrigation flows. It can be formed from many different materials, and is commercially available in metal and reinforced fiberglass with throat width ranging from 25 mm to 15 m (1 in. to 50 ft).

This type of flume is commonly installed in wastewater treatment plants for flow monitoring. The Parshall flume has performed well in these locations because of accurate results and low maintenance requirements. Because the channel is rectangular and a head drop of at least 70 mm (3 in.) is required for free flow, it is difficult to install a Parshall flume into existing sewers for I/I monitoring.

Palmer-Bowlus flume—The Palmer-Bowlus flume (Figure 3.4) is an excellent device for measuring flows in sewer systems. This open channel flume is easily adaptable to a wide variety of sewer diameters. However, a separate unit is required for each sewer diameter. The basic hydraulic principle is the same as that for a Parshall flume: constriction in the side and a step up in the sewer bottom causes the flow to go through critical depth. Critical depth can then be related to the flow rate. Development of a rating curve is shown by Metcalf and Eddy.[1]

The Palmer-Bowlus flume is available commercially in several materials and

FLOW MONITORING

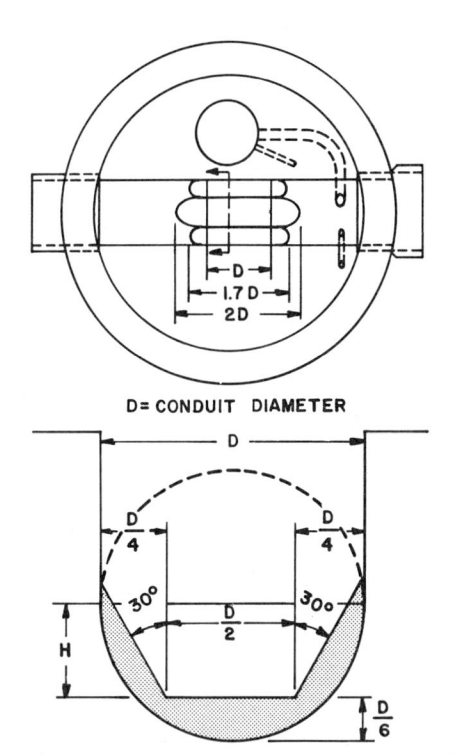

FIGURE 3.4. Palmer-Bowlus flume details.

configurations. Different types can be used for direct installation in sewers, or, for infiltration and inflow studies, in manholes. They can be temporary or permanent (Figure 3.5).

The main advantage of the Palmer-Bowlus flume is that it is easily adaptable to the circular section of sewer lines. It also requires a very minimal head loss, which prevents surcharging of sewers. The disadvantages are that the Palmer-Bowlus flume may not be as accurate as the Parshall flume and it usually has a low capacity.

Other flumes—The H-flume is useful for measuring flows over a very wide range. Parshall and Palmer-Bowlus flumes have a useful range of flow variations of about 10 to 1, but the H-flume can be used to measure flow variations of 100 to 1. This flume would be most convenient for measuring stormwater or combined sewer flows. The H-flume requires a free discharge at the outlet; therefore, a substantial head loss is required.

A trapezoidal flume is particularly suited for monitoring small flows. It has often been installed in earthen or concrete ditches. The flume channel has a flat bottom, which reduces the opportunity for silt accumulation, and the head loss is small.

Dye-dilution/chemical tracers. The dye-dilution technique is a simple, potentially accurate, and quick method for the determination of flows in sanitary sewers. Flows can be measured even under partially full or surcharged conditions without entering manholes. This method is normally employed to obtain instantaneous flow rates, but with added equipment can be used to monitor flows on a continuous basis.

Advantages—
• No entering of manholes;
• Saves time and provides instantaneous flow data on many sewer sections;
• Independent of sewer site, dimensions, velocities, and surcharging.

Disadvantages—
• Samples must be analyzed as soon as possible (most dyes decay in sunlight);
• Temperature corrections may be required;
• Instrumentation is expensive;
• Dye is expensive;
• Need at least 100 sewer diameters for dye mixing before sampling.

Chemical and radioactive tracers can be used to measure wastewater flows. The methodology utilized in chemical and radioactive tracers is similar to dye-dilution except the initial concentration of the wastewater stream must be determined. The flow in the wastewater stream is determined using Equation 1:

$$Q_s = \frac{Q_t (C_t - C)}{C - C_s} \quad (1)$$

where

Q_s = Stream discharge,
Q_t = Tracer discharge,
C_t = Initial concentration of tracer discharged,
C_t = Initial concentration of tracer in waste stream, and
C_s = Tracer concentration of sample.

49

EVALUATION/REHABILITATION

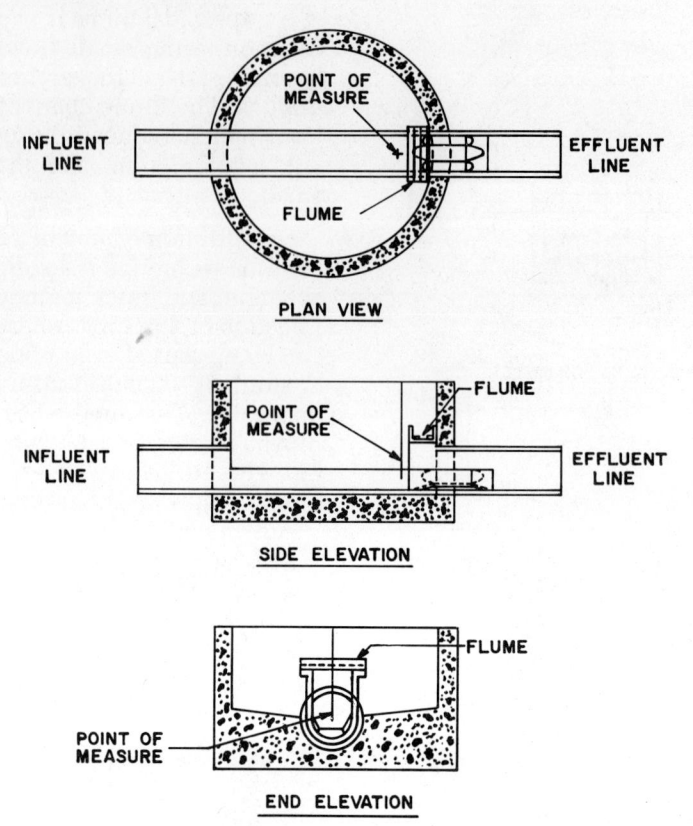

FIGURE 3.5. Palmer-Bowlus flume installation.

Dye, chemical, and radioactive tracers can also be used to measure velocity between two control locations. The time it takes for the center of gravity of an injected tracer to travel to a downstream control point is noted and the velocity is computed by dividing the distance between the control points by the travel time. Average depth of flow is then used to calculate flow.

Dye-dilution and chemical tracers, when conducted with care can provide flow data accurate to 5% or better.[2,3]

Velocity-area Method. Sanitary sewers generally operate only partially full and are thus classified as Open Channels. As described in Chapter 2, measurement of flow in the simplest terms takes the form of:

$$Q = AV \quad (2)$$

where all terms have been described previously.

Manning's equation—The familiar Manning equation for flow in open channel flow is shown below:

$$V = \frac{k}{n} R^{2/3} S^{1/2} \quad (3)$$

$$Q = \frac{k}{n} R^{2/3} S^{1/2} A \quad (4)$$

where

n = Coefficient of channel roughness,
R = Hydraulic radius (area divided by wetted perimeter),
S = Slope of hydraulic gradient (Note: S = slope of line only for uniform flow),
A = Cross-sectional area, and
k = 1.00 for metric units; 1.486 for English units.

There are three values that must be determined before a flow rate can be computed. These are depth of flow, pipe roughness, and energy gradient at the point depth is measured.

Not only must these values be determined, but their variability with relative depth must also be determined.[4] It is a common misconception that Manning's "n" is a constant. Actually, the roughness coefficient varies with liquid depth.

Calibrated discharge curves (stage discharge curves)—It is possible to utilize the depth recorder and obtain fairly accurate data by calibrating the flow monitoring location. This method relies on the depth recorder for the source of raw data and calibration of flows at various times and liquid levels using instantaneous weir measurements or other primary calibration devices. When utilizing this method, the depth is noted and a calibrated flow rate observed at the same time. By making observations at various liquid depths, a discharge curve can be developed to convert depth readings to calibrated flow rates.

Velocity. The direct measurement of velocity in a wastewater system can provide accurate and reliable flow data. A velocity reading and depth of flow can be used to calculate flow rates by using the continuity equation.

The average velocity can be readily determined by the one-point or two-point method. In the one-point method, the velocity meter is placed at the 0.6 depth and the measurement is used as the average velocity. The one-point method should be considered a rough approximation and used only where time or shallow depth of flow are significant factors. The two-point method records velocity at the 0.2 and 0.8 depths and the average of these two readings is taken to represent the average velocity.

Current meter—Portable current meters of the propeller type are commonly used to obtain velocities. Data are obtained from current meters by counting revolutions (then converting to velocity) or from meters that convert and directly display the measured velocity. Care must be taken when using current meters to ensure that suspended matter in the wastewater does not clog the meter, and that sufficient depth of flow is available to obtain accurate measurements.

Velocity probes—Portable velocity probes of the magnetic and Doppler ultrasonic types utilize no moving parts and are therefore not affected by clogging. Magnetic probes consist of wire coils that generate an electromagnetic field and electrodes that measure the induced voltage created when a conductor (wastewater) passes through the field. The induced voltage is converted to velocity. The probe is inserted into the flow to obtain measurements.

Doppler ultrasonic meters transmit a continuous high frequency pulse into the wastewater. A reflected pulse of different frequency is received and the difference between the two frequencies is proportional to the velocity. This in turn is displayed as a velocity measurement on the meter. Most ultrasonic probes are mounted on the outside of pipes making them ideally suited for obtaining velocities in force mains, sludge lines, and other pressure pipes. These devices can produce accuracy of 1 to 5%.[3]

A hot-wire anemometer can also be used for spot velocity checks. The passing fluid causes a heated wire or probe to be cooled. The temperature change is proportional to the velocity and can be recorded electronically. These devices are usually rather delicate, however, and are more suited for laboratory work than for measurement of flow in sewers.

Float measurements—Surface floats can be used to obtain velocity measurements in sanitary sewers. Velocities are calculated by timing a float between a known distance, usually two manholes. The distance traveled divided by the time determines the velocity. Surface floats record only surface velocities and results are considered approximations.

Pump station. Pump stations are a logical choice for monitoring wastewater flows

EVALUATION/REHABILITATION

that are because of location and inability to monitor upstream of the pump station to obtain accurate flows. It is important to calibrate each pump and possible combinations of pumps because the design capacity in most cases is not representative of actual field conditions. In a pump station with two constant speed pumps that alternate in operation and run together under high flow conditions, three calibrations will be required. (Note: This technique is applicable when the force main discharges freely and not into another common force main shared by other pumping stations. Possible pump combinations become excessive and difficult to monitor when other pump stations feed into a common, shared force main.)

Calibration using wet well drawdown/return—For this method it is necessary to obtain accurate dimensions of the wet well from as-built drawings or field measurements along with locations of incoming sanitary sewer lines. The following procedure is used to calibrate each pump and combinations of pumps:

• Familiarize yourself with the pump controls, the depth to which the wet well can be lowered without exposing the suction lines as well as the elevations of sewer lines entering the wet well.
• With all the pumps turned off and the wet well level below all incoming lines, measure the level of the liquid from some reference point.
• Start Pump 1 and let it run for 1 minute (shorter time if wet well empties too quickly), being careful to record time accurately.
• Record drawdown distance (D_d) from reference point at the time (T_d) the pump is turned off.
• With pumps off, record the time (T_r) it takes to refill the well to a known distance (D_r).
• Refill the wet well and repeat the procedure until three consecutive measurements are the same.
• If there is more than one pump, repeat procedure for each pump and possible combination of pumps. (For every instance, if a station has three pumps, calibration must be made on pump 1, pump 2, pump 3, pumps 1 and 2, pumps 2 and 3, pumps 1 and 3, and pumps 1, 2, and 3.)
• The pump flow rate can be calculated from Equation 5:

$$Q_p = \frac{(A_w)(D_d)}{T_d K} = \frac{(A_w)(D_r)}{T_r K} \quad (5)$$

where

Q_p = Calibrated flow rate for pump, m³/s(gpm),
A_w = Area of the wet well, m² (sq ft),
D_d = Depth of drawdown in time T_d, m (ft),
T_d = Time of draw down, s,
D_r = Depth of return in time T_r, m (ft),
T_r = Time of return, s, and
K = Conversion factor, 1 for metric units, 448.8 for cfs to gpm.

Inconsistent results most likely will be due to fluctuations in the incoming flow or infiltration directly into the wet well. Should this occur, it may be necessary to undertake calibrations at night during low flows, or temporarily plug the incoming line(s) or calibrate during low groundwater conditions. If the incoming lines are plugged, step 5 is omitted because no return flow will be measured and the effect of the incoming flow is zero.

Calibration using velocity meter—Velocity meters can be used to obtain the velocity in the force main while each pump, and combination, is calibrated. Sensors are placed on the discharge force main and velocity measurements taken. Care must be taken in locating the meter so that bends and valves do not cause inaccurate measurements. Conversion to flow rates are simply calculated using Q = AV, where the terms have been defined previously.

Calibration using discharge volume—For small pump stations the following procedure may be considered.

• Turn pumps off and plug incoming line

at the manhole downstream from force main discharge point.

- Turn pumps on and allow line to surcharge to depth D_o above outgoing pipe. (Care must be exercised not to cause backup into basements or homes.) Turn pump off.
- Turn on pump to be calibrated for about 1 minute, then turn off and record run time (T_r) and depth (D_1).
- Remove plug and allow surcharge level to decrease to a new D_o level.
- Calculate flow rate;

$$Q_p = \frac{A\,(D_o - D_1)}{T_r} \qquad (6)$$

where

Q_p = Flow rate of pump, m³/s (cfs),
A = Cross sectional area of manhole, m³/s (fps),
D_o = Initial liquid level prior to test, m (ft),
D_1 = Final liquid level, m (ft), and
T_r = Running time of pump, s.

Repeat to obtain consistent measurements.

Calibration using pump curve—The pump curve calibration involves plotting the system curve, describing the head loss through the discharge piping with respect to flow against the pump curve supplied by the manufacturer. The intersection of the two points gives the flow rate at which the pumping system operates. In actuality, the pump will operate along a small section of the pump curve as the flow is reduced when the wastewater level in the well drops. The system curve should be plotted for both high and low wet well levels (pump on and pump off).

Accuracy of this method depends on obtaining reliable values for the head loss through the system. Friction factors should take into account age, corrosion, and slime growths in the discharge and force main piping. Pump curves may also change slightly as the pump experiences wear. In general, this is not as accurate as some of the other calibration techniques described.

Summary—Once the pumps and combinations of pumps have been calibrated, it will be necessary to record the on/off cycles (events) of each pump. Many pump stations are equipped with meters on each pump that record the running time. This information is used for maintenance but can be recorded hourly or daily to obtain flows (that is, running time multiplied by the calibrated pumping rate is equal to the volume of wastewater pumped).

An inexpensive method to record running time is to install electric clocks on the motor circuits such that they operate only when the pump is on. The running time of the pumps can be recorded hourly or daily and converted to flows. It also will be necessary to install a clock that will operate only when both pumps are running in order to account for all possibilities.

Meters specifically designed for pump stations are available and will greatly reduce the time necessary in installation and monitoring. One event recorder can record on the same strip chart the time on, and time off of three (or more) pumps simultaneously. The strip chart provides a continuous record and will also show whether two (or more) pumps are operating simultaneously. Other solid state digital storage devices can store the same information in memory for easy retrieval without analyzing strip charts. Special event recorders are also available to record cycles on variable speed pumps.

Bucket test. The bucket test is probably the most widely used method to obtain accurate instantaneous flow rates for free falling flow. This technique is limited to small flows where the discharge can be contained in a calibrated vessel. To measure the instantaneous flow rate a bucket or container of known volume is placed in such a way that the entire flow is contained. A stop watch is used to measure the time it takes to fill the container. Once the container is filled, it is removed from the flow and the time of filling is recorded. The flow rate is calculated by simply dividing the measured liquid volume gathered by the elapsed time of filling. Several

EVALUATION/REHABILITATION

repetitions should be made to obtain consistent measurements.

Stage measurement. This is one type of device that can be inexpensive and can yield valuable information regarding extremes of flow depth. Cups are mounted on a vertical shaft at measured intervals and are used to approximate high water levels in a storm. Such cup gauges can be very useful during the early stages of a study program to define flows in the system and to determine the types of equipment required. They can also be used to supplement other equipment during the formal monitoring program, particularly if the budget is tight. Of course, such gauges must be checked and emptied after each storm or high flow period.

Automatic flow meters. Automatic flow meters continuously record various flow parameters with a minimum of labor. Data collected may be displayed, recorded on charts, stored on magnetic tapes or solid state memory, or even transmitted from the field to office by telephone or radio.

These meters can save considerable time and effort compared to manually recording flow data, but proper installation, calibration, and maintenance require individuals with basic knowledge of hydraulics and proper maintenance procedures for the meter in use. There are many automatic flow meters manufactured with various options and techniques for recording and analyzing flows. General capabilities of various automatic meters are discussed below.

Depth. Depth recorders are extremely versatile and can be used to measure liquid levels in a pipe, head over a weir, depth in a flume, or other applications where unattended depth measurements are necessary. Commonly used techniques to record liquid levels (depth) are: probe, bubbler, pressure sensor, float, ultrasonic, and capacitance/electronic.

Probe—A probe automatic recorder measures or senses the wastewater surface with a thin electrical wire. When the wire (probe) makes contact with the wastewater, a circuit is completed, and the probe stops and retracts slightly. Every few seconds the probe senses the liquid level. Data are transferred to the recorder either above ground or in the manhole depending on the equipment installation. It is necessary to clean and calibrate the recorder during installation and periodically during the monitoring period. This is accomplished by setting the recording device to the actual liquid level measured by a ruler or other device. After initial calibration, the probe and recorder will sense the liquid surface level and record the depth on charts or other permanent devices.

Bubbler—A bubbler recorder senses actual depth as compared to the probe, which senses a liquid surface level. A small pneumatic compressor, air tank, or canister supplies a low gas flow to the bubbler tube. The bubbler system then measures the water pressure at a selected point which in turn corresponds directly to the depth of water. A sensitive pressure transducer makes the comparison between the force necessary to expel a bubble from the sensor tip and the atmospheric pressure. This pressure differential is then converted to a depth measurement and recorded. Cleaning the tubes is necessary because the bubbler can clog with grease and other solids.

Pressure sensors—Pressure sensing devices may be mounted directly under water. These devices measure the fluid pressure directly on the sensing element and then electronically convert it to an equivalent depth of fluid and record it. Field cleaning and calibration checks should be considered standard procedure.

Float—The float recorder is a mechanical device, which, like the probe, measures the liquid surface level. A float in combination with a mechanical pivot or cable and pulley converts the vertical rise/fall of the liquid surface to a continuous depth recording. In some instances, a stilling well is fabricated because of turbulence in the flow. The float method requires accurate calibration of the initial depth at the time of installation with periodic depth checks during the monitoring period and possible cleaning. Data can be recorded on charts as depth measurements or directly converted to flow rates depending on the instrumentation.

Ultrasonic—The liquid surface level is measured with an ultrasonic device by transmitting a high frequency pulse and measuring the difference in time (or frequency) of the reflected return signal. A comparison is electronically made and converted to a depth (or, in the case of Doppler equipment, velocity) measurement. Measurement of the air-liquid interface requires an initial calibration of depth. Data are recorded on a continuous basis in the form of depth or flow rates. As with all the automatic depth recorders, ultrasonic devices can be used effectively in conjunction with weirs, flumes, or other primary measuring devices. Although a unit is not in contact with liquid or fouled by debris, waves, foam, and other floating materials can sometimes cause discrepancies in the results.

Capacitance/electronic—This type of recorder uses the flowing liquid as an electrolyte to sense the depth of flow. The probe can be inserted in a flume or in the invert of a sewer. Accumulation of grease and debris on the probe can cause false readings; therefore, field calibration checks and cleaning should be considered.

Velocity. Automatic flow monitors that use velocity measurements can provide accurate data even under highly fluctuating liquid levels. Velocity may be automatically recorded with ultrasonic Doppler methods, magnetic methods, mechanical current meters, or other methods. In most cases, a depth of flow is recorded along with the velocity in order to utilize the flow equation $Q = AV$ where the area component is a function of the liquid depth. The advantage of obtaining a velocity component is that even under surcharge conditions, accurate flow data can be obtained because the cross-sectional area becomes the area of the pipe being monitored. It is not uncommon for flows to reverse direction under surcharge conditions, and some velocity monitors will record a positive or negative velocity depending on direction. It should be noted that weirs, flumes, and Manning's Equation rely on open channel flow, and data should not be used when under surcharge conditions. These gaps in the data can be eliminated by using a velocity recorder, manually obtaining velocities during the surcharge condition, or installing meter pairs in consecutive manholes to determine the energy gradient.

Electromagnetic/Doppler meters—Velocity meters of the Doppler type are usually connected to the outside of the pipe to be monitored. This generally requires pipes that flow full (such as force mains) and have sufficient suspended solids for the transmitted signal to reflect back to a receiver. Absence of sufficient solids may be overcome by a bubble injection system; however, this generally is not required when measuring most wastewater. An advantage of ultrasonic meters is the ability to record flows in closed pipes without obstruction to the flow.

Orifice, nozzle, and Venturi meters—These three types of flow meters are used for measuring flow in completely filled pipes. The basic concept is to form a constriction in the flow so that the velocity increases and the pressure decreases.

There are many different configurations for this type of meter. It can be installed in the length of pipe or at the discharge end. The main disadvantage to this variety of meter is that the constriction provides an opportunity for solids to accumulate. In this respect, the nozzle or venturi meter is preferable to an orifice meter. In addition, grease can interfere with proper operation of the pressure taps.

Current meters and velocity probes—The use of current meters and probes is discussed earlier in the text.

Groundwater measurement. Groundwater levels are good indicators of potential infiltration into sanitary sewers. As such, groundwater levels should be measured as part of any monitoring program to help correlate and analyze flows.

Chapter 2 discusses the installation of

EVALUATION/REHABILITATION

groundwater gauges in detail. In certain instances, however, existing wells in the vicinity may be used to observe groundwater levels. These wells may be water supply wells, dewatering wells, or cooling water wells located in the study area. These wells may have long-term records of water levels that may be obtained from the owners.

The number of groundwater gauges used must be a function of the budget, watershed size and configuration, soil types in watershed, and similar factors. There should be enough gauges to allow the engineer to understand the basic groundwater level relationship to sewer elevation.

Rainfall measurement. Numerous rainfall measurement devices are available and should be considered as an integral part of any sewer system study where infiltration and/or inflow information is being analyzed.

It should be noted that site selection for a rain gauge is critical. Improper gauge location can cause errors in the data collected. Site selection should be done only after consulting a text on hydrology concerning rain gauge installation. The number of gauges required will be a function of watershed size, rainfall variation across the basin, prevailing storm travel patterns, and so on.

DESIGN OF A SYSTEM MONITORING PROGRAM

The decision to conduct system monitoring is based on a need for information, but the complexity of a program should only reflect how much information is needed to accomplish the job. Before determining the scope, the team designing a program should review the following questions:

- How much data are required?
- How accurate must the data be to satisfy the goals?
- Can the required data be found in existing files?
- What type of system is to be monitored?
- How experienced and knowledgeable are the personnel?
- What equipment is available?
- In what condition is the equipment?
- How much flow monitoring has been done previously on the same system?
- Are data from previous work applicable to the proposed program?
- What is the project duration?

These questions should be posed frequently both before and during the monitoring effort in order to keep the program goals clearly defined.

Research existing information. As in many other things, a good "foundation" or lack of it can mean success or failure. The foundation in this case is the research conducted prior to designing the flow-monitoring program. Indeed, much has been said in this chapter concerning research needs prior to monitoring. Some of these items are emphasized here.

First, the knowledge of the maintenance staff should be utilized as fully as possible. Experiences of people working on a system several years through a varying climate and changing conditions can often be used to identify specific locations where flow conditions are causing problems. Frequently, the most expensive short-term monitoring program does not locate such items.

Pump records in the system should be reviewed. Often such records are only in terms of hours of pumping as documented manually by maintenance personnel. In such cases, calibration of the pumps is always necessary to determine the true capacity of the pumps. Ideally, all pumps in the system are less than 10 years old and have well-calibrated flow meters and recorders on the discharge force main. This type of situation provides a consistent, long-term data base, if recorder charts are saved and filed properly.

Another valuable source of flow related data is the wastewater treatment plant. In addition to reviewing the flow records at the plant, the researcher should also secure the influent BOD and SS concentrations to get an idea of the extent of dilution of the flow at various times. Treatment plant personnel should be interviewed for two purposes: 1) to ask about flow patterns and general history; and 2) to establish a rapport for coordination during the monitoring program (that is, specific rates of pumping may be necessary during monitoring).

One data base frequently neglected is the long-term history of water and wastewater flows in the area. Simply plotting the monthly average water and wastewater flow values parallel to each other on the same time scale for a few years can reveal much about historical flows in the system. For example, a long-term increase in water production may indicate an increase in population or a change in water consumption habits; it can also mean that the water system is getting old and leakage is becoming a major problem. Also, a steady increase in wastewater with no accompanying increase in water production can indicate a deteriorating sewer system subject to more and more I/I. It can also mean a few "wet" years with above average rainfall.

When other long-term plots, such as precipitation, temperature, and BOD are added, this data source can be quite informative. It should be analyzed very early in the program.

When maps and other physical information about the system are secured and reviewed with the maintenance staff, copies should be available for marking to indicate problem areas experienced by the staff. Other data sources for the monitoring program are:

- Soil borings from projects in the area indicating groundwater levels.
- Geological information suggesting perched water tables, among other items.
- Older topographic maps before the sewer system was established, indicating natural drainage patterns, old swamps, old stream beds long since filled in during development.
- Storm sewer maps showing the density of the drainage system and areas with inadequate storm drains. Sanitary sewers in such areas are particularly subject to I/I.
- Design drawings showing trunk sewer profiles, if available, be pieced together.

Site selection. Once the researcher has exhausted the existing information on the sewer system, these data should be catalogued and reviewed thoroughly. If the existing flow data are insufficient, it is then necessary to begin designing the monitoring program.

At this point, the type of system being monitored dictates the type of equipment and procedures used. A large urban combined sewer system requires a completely different approach than the small suburban subdivision. Thus, the site selection should begin with the size of the pipes and the system in mind.

Key manholes. The sewer system should be divided into subsystems. The size of the subsystem is a variable but should generally include at least 5% of the system. The subsystem size, however, depends on the size of the system being studied.

The exact size of the subsystems is then determined when the key manholes are located. Manholes where the long-term (control) monitoring will take place should be selected with care. These manholes should be: on straight sections of pipe with consistent slope upstream and downstream; cleaned along with the sewer; large enough for easy access; relatively dry; free or relatively free from surcharging; located in highways or streets free of heavy traffic; and otherwise accessible. Finding such manholes can often prove difficult, necessitating much compromise with size of subsystems.

If information is available, it is often helpful to divide subsystems according to the way the sewers were constructed. All sewers in one subdivision should be isolated in one subsystem. Such a pattern can be used to find whether sewers in one subdivision are more susceptible to I/I than those in another. Otherwise, the selection of key manholes is left to the judgment of the program planners.

EVALUATION/REHABILITATION

Once the key manholes and major subsystems are located, it may be necessary to subdivide each of these subsystems into smaller areas to provide more useful and meaningful data. An economic evaluation of additional monitoring cost versus costs of performing other investigations, as well as the magnitude of the problems in an area, will give a guide to the length of these smaller subsystems.

Bypasses. If not identified at the beginning of the program, bypasses can cause many data interpretation problems after monitoring is completed. If the area takes in more than one sewer district, it is possible that upstream districts may be bypassing flow to a stream or another sewer system. Other bypassing is typical in an older system where newer trunk sewers were installed to intercept flow from certain points in the older system. If certain of these points are not documented, they can be overlooked in a later monitoring program. For best results, the program should include the monitoring of bypasses in order to establish total flow within the system.

Overflows. In a combined system, one is certain to find numerous combined sewer overflows. Also, in older sanitary sewer areas where drainage is insufficient, overflows may have been installed to relieve flooding in the system. In any case, if inflow is a concern in the monitoring program, significant overflows must be monitored.

Overflows may direct flow to waterways, to other subsystems, or to storm sewers. They are sometimes difficult to access, and frequently it is quite difficult to monitor flow through an overflow unless there is a manhole or a pump station on the overflow pipe.

Overflows need to be monitored in order to establish the total system flow if meaningful flow data are desired. Otherwise, the whole program could be rendered ineffective because of incomplete data.

Factors affecting flow. When observing flow in a sewer, the researcher should never assume theoretical conditions. Field conditions offer several variables, which must be uncovered and taken in account. For example, a manhole at a change of slope can cause distorted depth and velocity readings, especially if the change is great enough to cause turbulence at the manhole.

The type of monitoring equipment used will have distinct advantages and disadvantages under various field conditions. Equipment installation and monitoring results may be affected by some common conditions. For example, deposition in the invert of the sewer can cause distorted depth of flow data and can result in unnecessary turbulence and flow meter malfunction. Deposition should be removed through cleaning prior to the installation of the equipment.

A break in service connections in the section downstream from a monitor can cause slight surcharging in the key manhole, distorting flow data. Slipped joints can lead to deposition and surcharging similar to break in service connections. Dips in the sewer caused by settling or poor construction can also distort flow data. Broken pipe, depending on the severity of the break, can cause a number of problems. Minor cracks can allow significant infiltration into the sewer. Major collapses can cause flow stoppage.

The effects of the above conditions on the flow monitoring program can be minimized if the sewers up and down stream of the proposed key manholes are investigated. If any of these conditions are severe, consideration should be given to choosing another manhole.

Many other conditions can cause problems when monitoring. Water consumption is, of course, critical and should be analyzed to approximate wastewater flows. The local climate, season, and resultant precipitation or freezing, can upset the program.

One additional item that can upset flow monitoring is a tidal condition, if overflows to a tidal basin are being monitored. Depending on the situation, tides can cause reverse flows in the overflow pipes. Wind-induced currents on large bodies of water, such as lakes, can have a similar effect. These reverse flows obviously distort the flow picture.

Equipment selection and sizing. Monitoring programs and monitoring equipment are expensive and should be investigated thoroughly before purchase, lease, or use. Many meters require batteries. Bubblers require a supply of compressed air or nitrogen; tanks must be rented and refilled periodically. Certain spare parts such as pens and probes should be purchased along with the unit. It is recommended that extra meters be available and be used as exchange meters if a particular monitor becomes damaged or malfunctions. Otherwise, key data may not be obtained simply because of improper program planning. Other considerations are difficulty of calibration, susceptibility to moisture (or submerging), and sturdiness.

Most types of monitors are adaptable to large or small sewers and have been discussed previously. However, the manufacturer's specifications and recommendations should be reviewed in conjunction with the particular metering locations, conditions, and requirements.

Permanent and temporary siting. Two of the most critical factors in management of a flow-monitoring program are timing and data correlation. The only realistic method of ensuring that all flow-monitoring data can be correlated for the same intensity, duration, and frequency of storm, is that all monitoring sites measure simultaneously under the exact same storm conditions. This can best be accomplished by monitoring all sites at the same time. Even this may present correlation problems, however, because of geographical variations in rainfall and rainfall patterns.

During a system investigation, the flow monitor may be installed for a relatively short period of time as necessary. During a study, it is always best that all monitoring points be monitored at the same time. Sometimes, this may not be feasible because of the size of the system or other considerations. If different system (manhole) locations are monitored at different times, a data correlation problem can always be expected.

The most common method of trying to handle data correlation for different rainfall events is to provide a series of control monitors in the system. These control monitors should be installed on a permanent basis for at least the duration of the total flow-monitoring program in the collection system. A better situation would be to have them installed permanently in the system before a system study started and let them remain in the system. This will allow their use during the flow-monitoring period for data control and correlation of the effects of different storm events upon the system. It will also allow a control and check on any other work to be accomplished during other phases of a study program, and for the continuing system maintenance after any special, short-term study is completed. They must, however, be properly maintained and calibrated if their data are to be meaningful. Some of the uses of these permanent monitors are:

- Scheduling rehabilitation activites;
- Checking the effectiveness of rehabilitation;
- Scheduling maintenance;
- Determining infiltration and inflow trends; and
- Providing data concerning future area development.

Safety program. Safety is all too often played down or ignored as a nonproductive aspect of the flow-monitoring program. There are at least three reasons why it should be of the highest priority:

- 1. Proper safety can prevent loss of personnel time, which can cost money and, at worst, totally disrupt the program.
- 2. A strong safety program will convince the field personnel of the manager's concern for their welfare and will improve morale.
- 3. An inadequate safety program can be an invitation to a liability suit following an injury.

EVALUATION/REHABILITATION

An adequate safety program is essential. Of course, it is not adequate to merely provide personnel with equipment. They must be trained on the equipment and be constantly reminded of the need for safety.[5]

Monitoring necessities. Following research, site selection, adequate safety training and selection of equipment, there are a few additional items that should be considered before beginning the monitoring.

Maintenance of monitors. The most sophisticated equipment in the world will not yield consistent, accurate data unless it is regularly maintained both in and out of the manhole. When on the shelf, it should be dry, clean, and calibrated (if applicable). While being used in the manhole, it should be checked at least weekly and after every storm.

Regardless of the equipment selected, some degree of preventive maintenance or servicing will be required. This task may vary from cleaning debris deposited upstream of a weir to changing batteries or freon gas canisters. Most manufacturers and suppliers provide maintenance manuals for their equipment. The technical knowledge necessary to keep the equipment operational will vary greatly. Therefore, it is wise to establish the maintenance requirements when evaluating equipment. Maintenance logs should be kept on each meter and completed during field servicing of the meter. Information recorded will depend on the meter but may include the following:[6]

- Date, location, time, supervisor's name, serial number of monitor;
- Voltage on battery or changed battery;
- Flow conditions at time of check;
- Depth of flow at time of check;
- Velocity of flow at time of check;
- Notation of any equipment malfunction or replacement;
- Chart time versus actual time;
- Discrepancy in actual conditions versus measurements.

The maintenance log is a checklist to establish that all preventive maintenance work has been done and the monitors are set in the correct mode of operation.

Under all circumstances, if the field investigations listed above indicate that a monitor is operating improperly, it should be immediately repaired, recalibrated, or removed and another calibrated monitor installed in its place. In the latter case, recalibration can then be undertaken at a later time or at a more convenient location. Also, key manholes or monitoring points should always be checked immediately after storms to see that high flows have not disturbed the equipment or disturbed the calibration.

Quality control of monitoring data. Flow-monitoring data must be closely reviewed to obtain reliable results. Actual flow conditions recorded on the maintenance log must be compared with the recorded data to ensure that the flow monitor has recorded accurately. In addition, instantaneous flow measurements during the project using a primary device such as a weir should be undertaken and compared with recorded data. Factors that should be considered when reviewing flow data include:

- Did bypasses or overflows occur that will affect recorded data?
- Did maintenance logs show any malfunction of a unit? If so, how much of the data is salvageable and how much should be disregarded?
- Was the depth of flow or velocity ever outside the capabilities of the unit?
- If the collection system is surcharged, will this affect the data based on the type of equipment installed?
- Does the instantaneous flow measurement compare favorably with the recorded data?

A stage discharge curve may be developed for each key manhole before or during the monitoring. Depth and velocity readings should be taken to determine flow conditions (minimum flow to maximum flow) occurring in the sewer. These readings should be used to construct a plot of flow versus depth of flow. Depths on the charts can then be readily and accurately converted to flow.

To obtain the data necessary for a stage discharge curve, it may be necessary to mobilize the field personnel on a flexible-schedule basis. For example, to determine minimum flow, it is necessary to take readings between 1:00 a.m. and 6:00 a.m. or when investigations indicate minimum depth.

The use of computers has greatly improved the capabilities of data analysis; however, when a computer generates flow data, it is still extremely important to review the data for quality control and applicability.

Monitoring duration. The duration of the monitoring required at a particular manhole is defined by the data required. If it is only required to assess the dry weather flow pattern, only a week of monitoring may be necessary. If it is desirable to monitor particular flow conditions in the sewer, however, or to determine what flow is caused by a certain type of storm event, a long monitoring period of several weeks may be necessary. Even after that period, the budget for the program may be used up before "ideal" conditions are realized. The following comments about monitoring should prove useful to the program manager.

Instantaneous monitoring. Instantaneous monitoring, in many situations, can be used to gain valuable information. Chapter 2 discusses instantaneous flow-monitoring during flow isolation. Also, in some cases, instantaneous monitoring may be all that is available because of budget or time constraints.

It is useful to spot check for depth and velocity in the sewer at the start of the program to check dry weather flow and to get an idea of flow extremes. In combination with cup gauges, instantaneous depth/velocity checks can be used to define the range of flow in the sewer. The correct type of monitoring equipment then can be selected to match the flow conditions. As mentioned above, spot checks should definitely be used to recalibrate continuous monitoring equipment during the program.

One additional use of instantaneous depth/velocity measurements is for monitoring a single storm event. If sufficient personnel are available and sufficient prior warning of the storm event can be secured, teams of individuals can be stationed at several locations in time to "catch" the event. Spot depth/velocity measurements then can be taken at prearranged times (that is, on the half hour, or even every 15 minutes) during the storm event. Certain provisions must be made to ensure success with such a program. For example, personnel must be well-trained and available for the event; each team at a manhole must be equipped in advance; and equipment used by one team must be calibrated with equipment used by all other teams.

Random monitoring interval. Again, the monitoring interval need only be as long as necessary to monitor the desired flow conditions. A random interval program is frequently the result of a tight equipment budget. Then, because insufficient equipment is available to monitor flow in all key manholes over the same period of time, individual devices must be moved in such a way that equivalent conditions are monitored at all sites. The one-in-6-months storm may occur during the first week of monitoring at points 1, 2, 3, and 4. After moving equipment to points 5, 6, 7, and 8, however, an equivalent storm may not occur for months. It is necessary, therefore, that rain gauge and "permanent" monitors be installed so that some means of correlation may be determined. It should be noted that additional engineering time for correlation may offset some of the reduced cost of monitoring for this type of program. If it is necessary to monitor the 6 months storm flow at all locations, the budget should be flexible. If not, some compromise may be necessary regarding the desired data.

Continuous short-term monitoring. It is frequently desirable to monitor flow at each

EVALUATION/REHABILITATION

key manhole continuously for a short period to gather specific data for a study. Again, if the equipment budget is sufficient and the timing is right, all locations need to be monitored simultaneously for only a short time. In general, however, approximately 14 to 20 days should be considered a minimum duration. The actual monitoring time must be determined based on weather and rainfall conditions as well as data needs. Sometimes a monitoring duration of several weeks or months is required.

Continuous long-term or permanent monitoring. There are occasions when it is necessary to establish semipermanent or permanent monitoring at a specific site. For instance, a special district may be discharging to another district, or it may be desirable to monitor effectiveness of flow reduction during a large scale rehabilitation program; determine collection system capacity data, infiltration and inflow trends, schedule system maintenance; or investigate system expansion. In such cases, equipment costs may be significant and it may be necessary to modify an existing manhole or even build a new chamber especially suited for the new equipment.

The need for evaluating the equipment prior to purchase cannot be overemphasized, especially for a permanent monitoring site. The literature should be reviewed and other municipalities should be consulted. Vendors should be brought in and lists of installations secured. Several installations should be contacted to check reliability of the equipment. Nearby installations should be visited if possible to compare conditions with those at the proposed site.

REFERENCES

1. Metcalf & Eddy, Inc., Wastewater Engineering: Collection and Pumping of Wastewater." McGraw-Hill, Inc., New York (1981).
2. Kulin, G., and Compton, P.R., "A Guide to Methods and Standards for the Measurement of Water Flow." NBS Special Publication 421, Institute for Commerce, Washington, D.C., May, 1975.
3. "A Guide for Collection, Analysis and Use of Urban Stormwater Data," American Society of Civil Engineers, New York, (1976).
4. Gutierrez, A.F., and Siu, M., "Flow Measurement in Sewer Lines by the Dye-Dilution Method." Presented at the Texas Section, American Society of Civil Engineers 1982 Spring Meeting at Fort Worth, Texas (March 26, 1982).
5. "Safety in Wastewater Worker." Water Poll. Control Fed., Manual of Practice No. 1, Washington, D.C. (1975).
6. "Plant Maintenance Program." Water Poll. Control Fed., Manual of Practice No. OM-3, Washington, D.C. (1982).

Chapter 4
METHODS OF SEWER REHABILITATION

63	General Considerations		Inversion Lining
64	Pipeline Rehabilitation	75	Manhole Rehabilitation
	Root Control and Removal		Frame and Cover Rehabilitation
	Excavation and Replacement		Sidewall and Base
	Point Repairs		Rehabilitation
	Chemical Grouting	78	Service Connection Rehabilitation
	Linings and Coatings		Rehabilitation Methods
	Sliplining	81	References

Over the past two decades a number of methods have been developed and used to repair wastewater collection systems. When not limited by the structural integrity of the system components or the need to increase the capacity of the existing system, these rehabilitation methods can be successfully employed to restore use without excavating and replacing large portions of the system. Repair methods generally cost less than replacement, and most methods minimize open-trench excavation, resulting in minimal traffic disruption and public inconvenience. Sealing and repair techniques that have been used with success include: chemical grouts for pipelines and manholes, coating systems for manholes and large diameter pipelines, liners for pipelines, and (in some instances) manholes and point repairs.

The choice of method or combination of methods depends on the physical condition of the sewer system components (pipeline segments, manholes, and service connections)[1] and the nature of the problem(s) to be solved. Although this manual is directed to correcting conditions that allow I/I, the rehabilitation methods presented in this chapter also are applicable to other maintenance and structural problems, such as root intrusion and interior corrosion caused by the presence of sulfides.

The section in this chapter on excavation and replacement is limited to a discussion of the normal applications, advantages, and limitations of this rehabilitation method.

GENERAL CONSIDERATIONS

The following factors should be considered when selecting a particular rehabilitation method:

• Cost;
• Potential I/I migration;
• Traffic disruption and the potential for possible interference with other public utilities;
• Maintaining wastewater flow;
• Safety;
• Structural features;
• Other planned projects or identified needs;
• The long-term effectiveness of the rehabilitation method.

Until now, the initial costs of the repair methods had been given equal weight on the assumption that all methods have equal long-term effectiveness and service lives. Unfortunately, because most of the repair methods have been developed only during

EVALUATION/REHABILITATION

the last 20 years, there is insufficient long-term data to make definitive comparisons. The designer should make his or her own assessment based on the best information available.

The potential for groundwater migration must also be considered when evaluating all rehabilitation methods. For example, sealing portions of a manhole-to-manhole sewer section to minimize groundwater entry may cause the water table to rise and increase the hydrostatic pressure on all of the pipeline joints and cause other joints to begin leaking. Sealing only a few joints in a manhole-to-manhole section may simply transfer the points of groundwater entry.

Furthermore, sealing only the sewer main pipeline—by any of the methods discussed in this chapter—may transfer groundwater entry to the service laterals. There is also concern that, because most sewers are backfilled with granular materials that readily transfer water, sealing only isolated manhole-to-manhole segments in an area may transfer groundwater to other unsealed manhole-to-manhole segments located above or below the section sealed. Because of these and similar concerns, many programs in use today are designed to correct most of the identifiable defects in an entire drainage area, rather than to repair only selected points.

Although various rehabilitation techniques may be discussed in generalities, each rehabilitation project may present unique concerns and constraints that must be analyzed to ensure a cost-effective and successful program.

PIPELINE REHABILITATION

Pipelines can be repaired by both internal and external methods, although internal methods are more effective for most problems. Several internal rehabilitation methods are available to restore use to a pipeline without excavating and replacing large portions of the system. Some of these internal methods can be used for restoring structural integrity in addition to reducing infiltration and/or inflow.

The major disadvantage of most of the available internal methods is that they reduce the cross-sectional area of the pipeline. The amount of reduction is dependent on the method and the internal condition of the pipeline. Grouts can function by sealing a pipe joint internally as well as externally. As such, as an internal sealant the cross-sectional area of the pipe is not significantly decreased.

For internal rehabilitation techniques, the pipeline must be cleaned first. The cleaning requirements vary somewhat for the different rehabilitation techniques, the general cleaning operation common to all techniques is described in Chapter 2 of "Operation and Maintenance of Wastewater Collection Systems."[2] In addition, any roots that have intruded into the pipeline must be removed before an internal rehabilitation method can be used. Root removal also minimizes maintenance problems.

External methods, other than excavation and replacement, can be used to stabilize soils, fill underground voids or washouts, and reduce groundwater movement. The two primary methods available for these purposes are chemical grouting and cement grouting. External grouting is accomplished by pressure-injecting the grout through a pipe into the underground soils. The injection pipe is placed in a predrilled hole, adjacent to the pipeline to be grouted. Generally, chemical grouts are used for fine soils and cement grouts for medium sands or coarser soils because of the larger size of the cement particles. Except for soil stabilization, however, these external methods are not generally effective for solving pipeline problems. Thus, further discussion in this chapter is limited to other methods.

Root control and removal. Removal and control of roots that intrude into sanitary sewer systems is an ongoing maintenance task for most system operators. An effective root-control program requires an understanding of the basics of root growth.

Roots grow toward moisture by a continuous process that adds cells, one at a

REHABILITATION METHODS

time, at the end of the root. This cell-by-cell growth enables roots to penetrate small openings in sewer pipes. Because roots seek moisture, root intrusion tends to be a problem only in sewers installed in soils where moisture is limited. This includes pipelines that are continuously or seasonally above the groundwater table. Root intrusion is a more common problem in service connections because they are normally shallower and frequently not as well constructed as the sewer mains. Figure 4.1 shows how severe this problem can become.

Root intrusion not only may increase the rate of infiltration into the sewer pipe or joint by expanding the opening, it may also substantially increase the potential for pipeline blockage. Once a root enters a sewer pipe, it expands to increase the mass available for moisture absorption. This not only screens out some of the solids in the wastewater but decreases upstream velocities, intensifying problems with deposition of solids in the sewer.

The most effective root control method is to prevent roots from entering the sewers in the first place. Installing watertight lines that are free from imperfections and will not crack, break, or deteriorate during service is the ideal solution. This may require that materials and construction methods meet or exceed current standards. It also may necessitate increasing onsite inspection during the installation of pipelines and service connections. The potential for root intrusion also can be reduced by discouraging tree planting near sewerlines.

Once roots have penetrated the pipeline, controlling the problem in general consists of mechanical removal, chemical treatment, or both. Typically, maintenance crews cut the roots using a sewer rod and an auger tool. Although this corrects the immediate blockage problem, it can actually increase root growth over the long term because roots will frequently grow back in a thicker mass after each cutting operation. Because of this, root cutting should be followed by chemical treatment or by flooding the pipeline with scalding water to retard root regrowth.

Various herbicides are available for chemical treatment. Chapter 5 describes these materials and their characteristics. Techniques for applying herbicides generally consist of flood treatment (soaking), spraying, or foaming. The flood treatment method permits easy control of chemical concentration and contact time during application, ensures that the herbicide penetrates the roots, and may kill some roots outside the pipe. The major disadvantage of this method is that it interrupts sewer service for about 1 hour.

Herbicides can be sprayed on the roots with standard hydraulic cleaning equipment. The short contact time between herbicide and root for this method, however, has not yielded significant long-term root control.

The use of foam herbicides is increasing because they are relatively easy to apply. In addition, foam is effective in treating root intrusion in building services without the problems generally associated with the flood treatment technique.

Chemical grouts used to seal infiltration sources also may be supplemented with herbicides to control root intrusion. An

FIGURE 4.1. Section removed from a sewer pipeline showing a severe roof intrusion problem.

65

EVALUATION/REHABILITATION

advantage of using grouting materials with herbicides is that it addresses two problems—infiltration and root intrusion—using the same rehabilitation technique. If the pipeline is already subject to moderate or heavy root intrusion, however, the roots must be removed before the chemical grouting is applied.

Additional information is available regarding the causes of root intrusion and various control techniques.[3, 4]

Excavation and replacement. Excavation and replacement of defective segments of a sanitary sewer system is an effective method for minimizing I/I. This method also is used to correct structural defects and alignment problems in existing pipelines. Unfortunately, it is usually the most costly method and causes the most inconvenience to the public.

Excavation and replacement of defective pipe segments is normally undertaken when the structural integrity of the pipe has deteriorated so severely, for example when pieces of pipe are missing, pipe is crushed or collapsed, or the pipe has large cracks—especially longitudinal cracks, that alternative rehabilitative techniques are not feasible. In addition, pipeline replacement is often required when the pipe is significantly misaligned.

Pipeline replacement provides an opportunity to correct misalignment of line or grade, increase the hydraulic capacity of a particular segment, repair improper or poorly constructed service connections, and eliminate direct sources of stormwater entry such as catch basins and area drains. Replacing pipelines can also remove "incidental" I/I sources, such as relatively minor joint leaks that, individually, are not cost-effective to remove.

Because this method replaces outdated pipeline with modern construction materials, it increases the service life of the pipeline. A useful service life of over 50 years should be provided with the materials and construction techniques currently used. These include the several types of pipes, pipe joints, and manhole construction materials currently available.

Chapter 5 describes these materials. A discussion of current construction techniques is presented elsewhere.[1]

The primary disadvantage of pipeline replacement is the high cost. Analyses to determine the cost-effectiveness of pipe replacement must include all costs associated with the replacement. These costs typically include pavement removal and replacement, excavation, possible substitution of select backfill to replace poor quality existing material, dewatering and shoring, pipe materials and couplings, and traffic control. Potential cost increases resulting from interference with other underground utilities and narrow easements or limited space for construction also must be considered. In addition, consideration must be given to the need for temporary flow rerouting to maintain sewer service to upstream connections and for special measures needed to minimize interruption of service along the affected line segment. Depending on the service life assumed for other rehabilitation methods, the possible higher capital costs may be somewhat offset by the longer service life a new line provides.

Excavation and replacement is the primary method currently used to correct defective service connections. Factors that influence the costs of replacing service connections include those discussed above, as well as access to the service connection, legal considerations of work on private property, jurisdictional considerations concerning ownership of the connection to the lateral sewer, and the location of the service connection to the sewer pipeline, that is, costs are increased if the sewer main is located in the parkway on the opposite side of the street). Restoration of landscaping or other site improvements can significantly increase the cost of sewer replacement on private property.

Point repairs. Point repairs may be used to correct isolated or severe problems in a pipeline segment, they can be used to totally correct the defects within a line segment, or they can be an initial step in the use of other methods. Point repairs are

REHABILITATION METHODS

similar to excavation and replacement, because most point repairs require excavation and usually some replacement. Point repairs, however, usually are limited to the replacement of only a short portion of a pipeline or service connection, instead of the entire length.

The technical considerations and the factors influencing the cost of a point repair are the same as those described above for excavation and replacement. When making point repairs, however, special consideration should be given to the materials and methods to be used to connect the replacement pipe to the existing pipeline. Flexible couplings are often used to join the pipes together. Concrete collars and encasement also are used occasionally.

The repair of pipe connections to manholes is another type of point repair. This is covered later in the "Manhole Rehabilitation" section of this chapter.

Chemical grouting. Chemical grouting is the most commonly used method for sealing leaking joints in structurally sound sewer pipes. Using special techniques and tools, chemical grouts can be applied to pipeline joints, manhole walls, wet wells in pump stations, and other leaking structures. Small holes and radial cracks may also be sealed by chemical grouting.

Advantages. Chemical grouting can be less expensive initially than other options for sewer rehabilitation. Generally, sealing a line segment usually takes only a few hours, so that, with proper scheduling, chemical grouting can be accomplished with minimal traffic interruption. Figure 4.2 shows a leaking sewer joint being sealed with chemical grout.

Because chemical grout is applied internally, this method does not damage or interfere with other underground utilities such as gas lines, water lines, or telephone

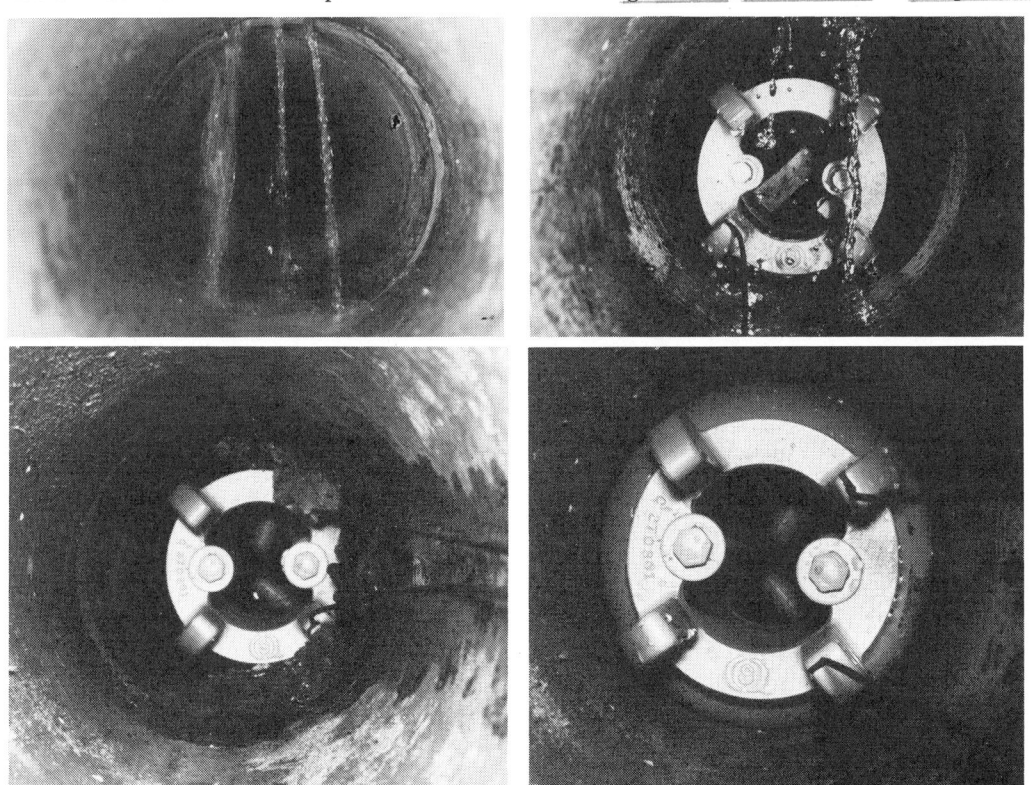

FIGURE 4.2. Chemical grout sealing.

EVALUATION/REHABILITATION

cables. All of the chemical sealant is applied with special equipment inside the pipe; therefore, no excavation is required. It also eliminates the need for surface restoration, such as pavement or sidewalk replacement and ground cover reseeding.

Limitations. Chemical grouting does not improve the structural strength of a pipeline. For this reason, chemical grouting should not be considered when the pipe is severely cracked, crushed, or badly broken. Also, when applying chemical grout to a sewer pipeline, the potential for dehydration should be considered; dehydration may reduce the service life of the grout. Chemical grouts, once applied, may dehydrate and shrink if the groundwater drops below the pipeline and the moisture content of the surrounding soil is reduced significantly. The amount and duration of the shrinkage depends on several factors such as the length of time the groundwater remains below the pipe, the moisture content of the soil while the groundwater level is depressed, the soil type, the type of grout, and the additives included in the grout to control shrinkage.

Some joints and cracks may be difficult to seal chemically using gel grouts when large voids exist outside the pipe joint. In this instance, extremely large quantities of grout may be required to seal the joint, if sealing is possible at all. Other joints cannot be sealed with either gel or foam grouts because they are badly offset or misaligned. Offset joints may prevent the inflatable rubber sleeves of the sealing unit from seating properly against the walls of the pipe, making it impossible to isolate and seal the joint. Chemical grouting of longitudinal cracks in pipe also is not generally feasible because the grout may flow through the crack past the sealing equipment and leak back into the sewer.

Materials. Acrylamide gel, acrylate gel, urethane gel, and polyurethane foam are the main types of chemical grouts currently available. The material characteristics of these sealants are presented in Chapter 5.

Application techniques. The chemicals necessary to form acrylamide or acrylate gels are usually mixed in two separate tanks and pumped through separate hoses to the pipeline joint to be sealed. One tank is used to mix and dispense a solution of water and the acrylamide or acrylate grouting material. The other tank is used to mix and dispense a solution of water and a catalyst. The water and catalyst solution initiates the chemical reaction when mixed with the acrylamide solution. Additives can be included in either solution to help control shrinkage, reaction or "gel" time, and other variables.

From the mixing tanks, the two solutions are pumped through separate hoses to the point to be sealed. The solutions are mixed as they are pressure-injected into the leaking opening, initiating the chemical reaction. This reaction changes the two solutions into a gel almost instantly when the predetermined reaction time has elapsed. The gel time can be controlled from just a few seconds to several minutes. The acrylamide and acrylate gels stabilize the soil around the defective joints or cracks by filling the voids.

Urethane grout materials may be used to form either an elastomeric gel, much like the acrylamide and acrylate gels, or a rubber-like foam. Urethane gel is different from the acrylamide or acrylate gels in that water is the catalyst for the urethane gel material. The properties of the reacted gels are also different (see Chapter 5). Urethane gel seals pipeline joints by forming an elastomeric collar within the pipe joint as well as by consolidating soils and filling voids outside the joint.

Urethane gel is applied in essentially the same manner as the acrylamide and acrylate gels; however, the equipment used for the polyurethane foam grout, which also is activated by water, is normally used. Different equipment is normally used because it is very difficult to remove all traces of the water solutions from the mixing tanks and the chemical feed lines. Any water remaining prematurely triggers the chemical reaction, clogging the equipment, or feed lines, or both.

REHABILITATION METHODS

Polyurethane foam differs from the gel grouts in that the foam is used to form an in-place pipeline gasket. The application techniques and equipment (discussed later in this chapter) used for the placement of the foam grout do not fill voids or stabilize the soil outside of the pipe joint.

Grouting small- and medium-sized pipes. Small and medium diameter pipes can be sealed using a hollow metal cylinder with inflatable rubber sleeves on each end of a center band, called a "packer." The center band on some packers is also inflatable. The type of packer depends on the type of grout being used. An inflated packer can be used both to test and chemically seal a pipeline joint.

Typically, a van is used as the operation and control center for a TV monitor, pumps, air compressors, and the chemical feed system equipment used for this method. A closed-circuit TV camera is used to position the packer over pipeline joints and radial cracks for testing or sealing. As shown in Figure 4.3, both the packer and the TV camera used with this method are pulled along with a cable from manhole to manhole, and the entire process can be viewed on the TV monitoring screen within the van.

The test-and-seal technique is used on most pipeline segments selected for sealing. This technique minimizes the potential for groundwater migration to other, unsealed joints in a manhole-to-manhole sewer segment. With this method, each joint is first tested to determine if it is airtight.

After the joint has been located with a closed-circuit TV inspection camera, the rubber sleeves at each end of the packer are inflated until they form a seal against the inside of the pipe and isolate the joint to be tested. If the joint fails the air test, the grout and catalyst solutions are pumped under pressure into the void between the two inflated sleeves and out into the surrounding soil.

The amount of grout needed to seal a defect depends upon the size of the leak. The acrylamide, acrylate, and urethane gels usually are pumped until the gel solidifies; the back pressure will then indicate to the operator that the leak has been successfully sealed. The operator then deflates the rubber sleeves and moves to the next joint for testing and possible sealing and continues until all the joints and radial cracks in the manhole-to-manhole segment have been tested and, where necessary, sealed.

The success of grouting using a packer unit can be assessed by conducting a low-pressure air test of the sealed defect once the setting time has elapsed. Pressures must be controlled to avoid rupturing the grout seal.

Large diameter pipe grouting. For grouting large diameter pipes, pressure grouting or manual placement of oakum soaked with grout may be used. The pressure grouting may be accomplished using either pipe grouting rings or predrilled injection holes.

Grouting using sealing rings (Figure 4.4) requires the use of a small control panel, chemical and water pumps, and various other accessories depending on the type of sealing grout being used.

To seal joints using grout sealing rings, a worker must enter the line, manually place the ring over the joint that requires sealing, and then inflate the ring to isolate the joint. The sealing rings work on the same principle as small diameter pipe packers, that is, sealing grout is pumped into the small void between the pipe wall and the face of the ring. In this method, though, the grout is pumped through a hand-held probe. As the pressure in the

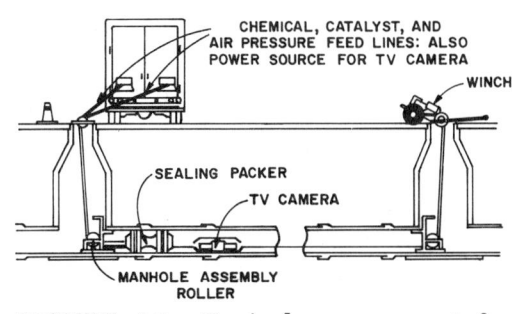

FIGURE 4.3. Typical arrangement for applying chemical grout to small diameter pipe.

EVALUATION/REHABILITATION

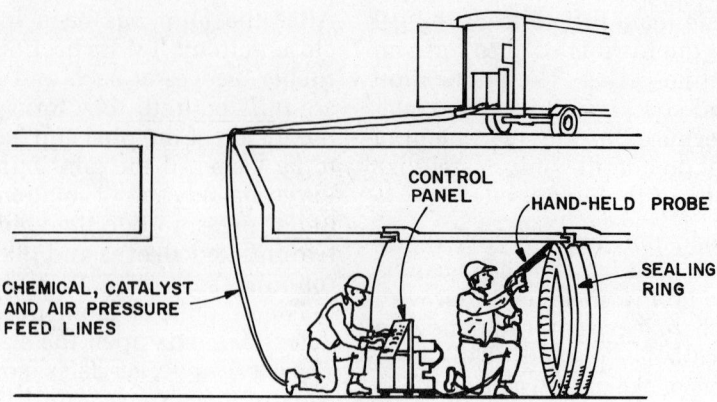

FIGURE 4.4. Typical arrangement for sealing large diameter pipe with grouting rings.

small void increases, the grout solution is forced into the joint and surrounding soil. After the catalyst solution is injected, the grout cures, sealing the joint from infiltration.

The probe injection method uses most of the same equipment as the sealing ring method, except that sealing rings are not required. As before, equipment depends on the type of grout being used. The worker enters the line to drill holes either into or next to the joint. The actual number of holes depends on factors such as size of pipe and length of crack. Once the holes have been drilled, one of the various techniques to apply the grout can be used. The sealing grout is injected by special injectors and equipment much like that used with the sealing ring. The injector is connected to this section of pipe through the wall and the chemical sealant is pumped into the joint.

A second method is to use a tapered probe at the end of the injector. The probe is manually pushed into the drilled holes and held in place during the pumping sequence. The grout and catalyst are pumped to the probe in separate hoses. Locating the holes near or at the joint enables the grout to enter the holes and fills any voids around the joint.

Cost and feasibility. Many factors can influence the cost-effectiveness of using chemical grouting to repair sewer pipelines. When considering this method, the pipeline should first be visually inspected to determine cleaning needs, the extent of root intrusions, and the structural integrity of the pipeline. For grouting to be effective, the pipeline must be relatively free of sand, sediment, and other deposits.

Roots in a pipeline can cause severe structural damage, making total replacement more attractive than sealing some of the joints and replacing other parts of the line where roots have damaged the pipe. Moreover, if joints are sealed where roots have previously entered, but the roots have not been effectively removed or inhibited, the root problem is likely to return.

The inside walls of the pipe must be smooth enough to allow proper seating of the packer. The packer can compensate for a certain amount of roughness, but it will be unable to seat properly if the pipe is badly eroded. If a line has several badly cracked or broken sections, partial replacement may first be necessary.

As previously discussed, potential groundwater migration must be considered when evaluating chemical grouting. The test-and-seal technique discussed minimizes the potential for groundwater migration to other, unsealed joints in a manhole-to-manhole sewer segment. Sealing only the sewer mains, however, may transfer groundwater entry to the service laterals. There is also concern that, because most sewers have been backfilled with granular material, which readily transfers water, sealing only isolated manhole-to-manhole segments in an area may transfer groundwater to other unsealed manhole-to-manhole segments.

The service life of the grout is an important consideration. Acrylamide grouts have been used successfully since the 1950s to stabilize soils and help control underground water movement in tunnels, dams, dikes, pits, and various other underground structures. The urethane grouts are more recent. Successful applications of the polyurethane foam have been in place since the early 1970s. Urethane gels have been successfully applied since 1980.

Pipe size, joint spacing, and the percentage of joints requiring sealing are factors to consider in determining the cost of chemically sealing a line. The larger the pipe, the higher the cost because of increased manpower, equipment, and material. The number of joints requirings sealing in a manhole-to-manhole sewer section is another significant factor to consider in estimating the sealing cost. The number of joints requiring sealing is usually determined by estimating the percentage of joints that will require sealing, multiplied by the manhole-to-manhole segment length and divided by the known joint spacing. Another factor in sealing larger lines is that daily production is lower and rerouting of wastewater flow around the section(s) being sealed often is required. Larger lines are also more difficult to clean in preparation for sealing.

Linings and coatings. Linings and coatings can be used to protect pipelines from internal corrosion.[5] Most linings, however, are integrated into the pipe when it is made and, thus, are not viable for rehabilitation of an existing system.

Reinforced shotcrete (gunite)—Gunite is a mixture of fine aggregate, cement, and water applied by air pressure using a cement ejector. Compared to cement mortar linings, gunite is denser and has a higher ultimate compressive strength. Like cement, gunite improves a pipeline's structural integrity. In fact, the greater the structural deterioration, the more effective the gunite process is compared to cement mortar linings. Gunite adheres well to other concrete and brick sewers and is more corrosion-resistant than normal concrete. Its finish, when troweled, is similar in smoothness to cement mortar linings, and it improves a pipeline's hydraulic characteristics.

Gunite is ideal for extremely deteriorated large sewers where persons and equipment can work without restriction. Long lengths of sewers may be effectively renewed with little excavation and minimal traffic disruption.

Installing gunite takes more time than installing a cement mortar lining. It can be applied under low wastewater flows; however, dewatering the pipeline totally is more effective. For pipelines carrying corrosive or aggristic wastewater, special aggregates and high alumina cements may be used. Welded wire mat or small diameter rod reinforcing is used for structural gunite applications.

Sliplining. Sliplining involves sliding a flexible liner pipe of slightly smaller diameter into an existing circular pipeline and then reconnecting the service connections to the new liner. Where applicable, sliplining an existing pipeline segment usually can be completed in less time and at a considerably lower cost than conventional excavation and sewer replacement.

Polyethylene is the most common material used for sliplining pipelines. Polyethylene pipe used for sliplining currently is available in nominal diameters ranging from 100 mm (4 in.) through 3000 mm (120 in.). A variety of other piping materials are used for sliplining. For a discussion of these materials and their characteristics, see Chapter 5.

Applications. Sliplining is used to rehabilitate extensively cracked sewer pipelines, especially lines in unstable soil conditions. It also is used to rehabilitate deteriorating pipe installed in a corrosive environment, pipes with massive and destructive root intrusion problems, and pipelines with relatively flat grades.

EVALUATION/REHABILITATION

One of the advantages of relining rather than replacing a sewer main is that it requires a minimal excavation, which limits traffic disruption and interferences with surface structures. This is especially important where a sewer main is not in an accessible right-of-way, such as those located under retaining walls, landscaping, or portions of buildings. Sliplining also can be used to avoid the extensive dewatering that is necessary for conventional open trench construction. With this rehabilitation method, dewatering needs are limited to the insertion pit and the service connections. The insertion of a pipe with heat-fused or, in some cases, gasket joints, will eliminate problems with root intrusion and joint infiltration. Another advantage is that this technique is actually a form of replacement since the liner is a new pressure-capable pipe itself.

Sliplinings can be installed in pipelines having moderate horizontal or vertical deflection. Where such deflection has been caused by shifting soils, sliplining has the advantage that a normal amount of future settlement or deflection can be accommodated easily with the flexible liner pipes.

If the existing sewer main joints are offset, service lateral taps are protruding, or if the diameter of the sewer has been significantly reduced in some other manner, the size of the liner pipe may have to be much smaller than the existing sewer main. Such conditions can limit the utility of this method, depending on the extent of the problem and the condition of the pipeline and service connections.

Installation procedure. Before installing a liner pipe, the existing sewer main should first be inspected by closed circuit TV to identify all obstructions such as displaced joints, crushed pipe, and protruding service laterals. The inspection also should locate all service connections that will need to be connected to the new liner pipe. Next, the existing pipe must be thoroughly cleaned. Depending on the obstructions identified, it may be necessary to proof test the existing pipe by pulling a short piece of liner pipe through the sewer section.

As illustrated in Figure 4.5, liner pipe can be pushed or pulled through an existing sewer main. The method chosen is partially dependent on the type of pipe to be inserted and the type of joint. Pulling is the most common method used and requires excavating a pit at one end of the section to be lined. It is usually not necessary to excavate a pit at the pulling end because most equipment fits into standard sized manholes. The excavation down to the springline of the existing pipe should have an entry slope grade of a minimum of 2.5 times the depth to the sewer invert. The length of the level section of the pit should be at least 12 times the diameter of the pipe being inserted. The width of the shaft can be as small as necessary, consistent with the diameter of the pipe, type of soil, height of water table, and other working conditions.

To pull a liner pipe through an existing sewer, a steel cable is threaded throughout the existing pipe and attached to a pulling head, which guides the pipe end past minor obstructions and prevents the entry of debris. It also may be necessary to put guards over the edges of the existing pipe at the inlet end to prevent damage to the pipe during the insertion procedure. Sections of the liner pipe that have been butt-fused together in the field into a flexible continuous pipe are then pulled through the sewer by use of a winch.

The pulling speed is unlikely to exceed about 300 mm/s (1 ft/sec), and slower speeds are necessary under more difficult conditions. The pulling operation tends to stretch the pipe, and excessive stretching (more than 1.5%) should be avoided. Stretching of about 1% of the total length pulled is common. This stretching will be recovered over a period of time about equal to the length of time it took to complete the pull. If the work is done during warm weather, an additional change in length may also be observed. A difference of as much as 1 mm/1 m x 5°C (1 in./100 ft x 10°F) can result before and after installation, which should be allowed for in the length of insertion pipe used.

Once the lining has stabilized in length,

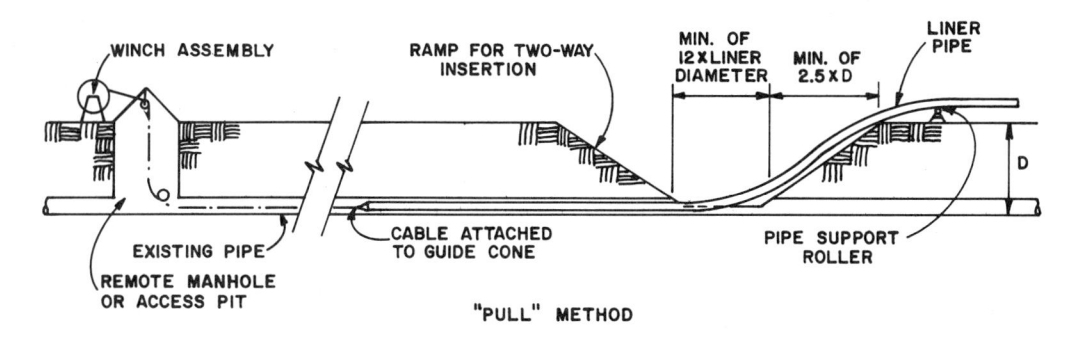

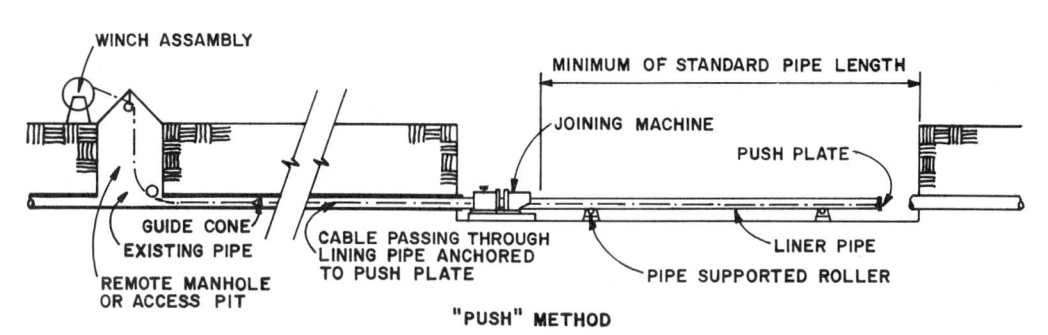

FIGURE 4.5. Sliplining installation methods.

further steps may be necessary to protect it. It is suggested that concrete anchor blocks be poured around the pipe at all points where the old pipe has been broken away, at excavated points, and at service connections. This is particularly important when excavations are shallow. Where the sliplining is smaller than two-thirds the inside diameter of the existing pipe, spot anchoring or continuous encasement by grouting the entire length is desirable to minimize potential movement of the sliplining caused by changes in the groundwater level or temperature.

There is some disagreement about whether the annular space between the new liner pipe and the old sewer should always be continuously encased by pressure grouting. There is also disagreement about this practice in the United Kingdom, as documented in the proceedings of an international conference held in 1981.[6] Some consideration should be given to grouting the annular space outside of the liner pipe to prevent movement.

Wastewater flow in the existing main may or may not need to be interrupted during insertion of a sliplining. Depending on the size of the insertion pipe, the flow may continue in the annular space between the two pipes. If not, it may be necessary to temporarily plug the upstream lines and pump the flow around the section being lined using above-ground piping.

Reconnecting service connections. Installation of a polyethylene liner requires multiple excavations to connect each existing service lateral to the new liner pipes. Various manufacturers have developed methods for connecting service laterals. These methods generally can be grouped into three areas: remote connector fittings, heat fusion saddles, and mechanically connected tapping saddles.

To use a remote connector, the service lateral is excavated and the existing service piping is broken away. A cutting tool is then inserted through this opening and a hole is cut through the liner pipe. Once

EVALUATION/REHABILITATION

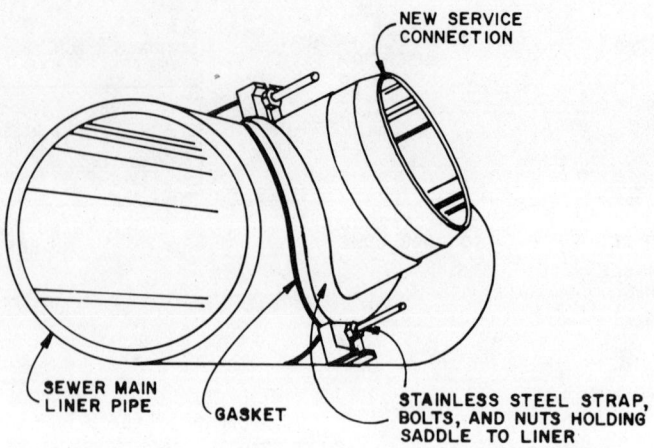

FIGURE 4.6. Full encirclement tapping saddle.

the hole is cut into the main, a piece of pipe is heat-fused to the connection end of the lateral pipe. A gasket on the connector fitting further minimizes potential infiltration. A special trimming tool then removes any excessive protrusion of the new polyethylene service connection.

Heat fusion saddles require excavating the service connection to the main. Portions of the old main and service connection are then broken away from the new liner and the area for the fusion of the saddle connection to the liner is cleaned. After the fusion has cooled and become firm, a hole is sawed through the outlet of the saddle fitting into the main's interior. The service line can then be connected to the saddle with flexible couplings and stainless steel straps.

The tapping saddle procedure involves a full-encirclement saddle fitting held in place by stainless steel straps (Figure 4.6). The old service connection is first removed from the sewer main. The surface is then cleaned and a hole is drilled in the new liner at the location of the connection. Neoprene gaskets are used between the underside of the saddle and the liner to provide a tight seal. The saddle is then fitted to the hole in the liner, and the straps are drawn around the saddle and the sewer main to complete the junction.

Inversion lining. A patented system is available through licensed contractors for installing liners from the ground surface using existing manholes for access. This system uses a flexible lining material that is thermally hardened. After the lining system has been installed and cured, a special cutting device is used with a closed-circuit TV camera to reopen service connections located with the camera before the liner was installed.

The liner is installed through a large tube that is placed into a manhole. At the bottom of the tube is an elbow that guides the liner into the pipe to be lined. The liner is inserted into the tube and attached to the end of the elbow (Figure 4.7). Then, water or air is pumped into the tube. As the pressure builds, the flexible fabric is pushed through the pipe and is inverted into place. The unique method by which this material is installed allows great flexibility in changes in liner bends.

With the liner in place, the water or air is heated to appropriate temperatures to cause the thermo-setting resins in the material to cure and harden. Following curing and cool-down, the water is drained away and the ends of the lining are cut and sealed at the two manholes. The characteristics of the lining material and the cured liner are discussed further in Chapter 5.

Applications. Because inversion lining can be accomplished relatively quickly and without excavation, this method is partic-

REHABILITATION METHODS

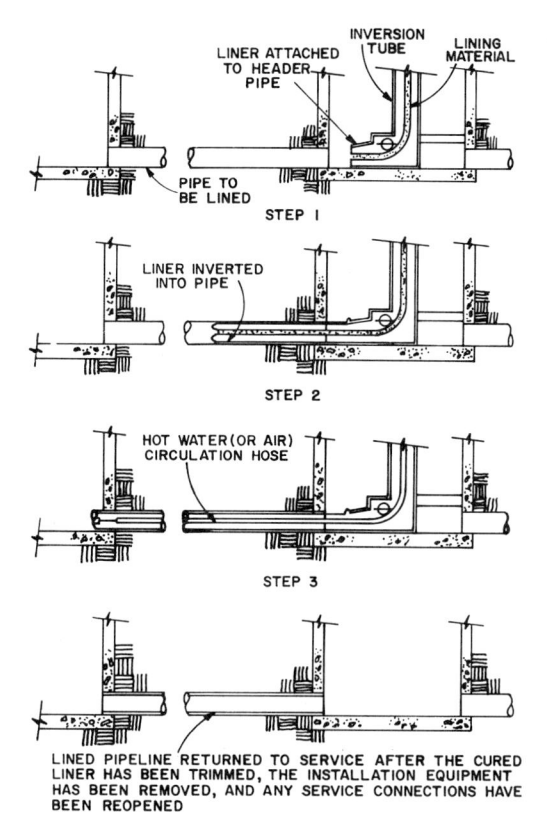

FIGURE 4.7. Inversion lining installation procedure.

Steps shown:
- STEP 1: Liner attached to header pipe; inversion tube; lining material; pipe to be lined.
- STEP 2: Liner inverted into pipe.
- STEP 3: Hot water (or air) circulation hose.
- STEP 4: Lined pipeline returned to service after the cured liner has been trimmed, the installation equipment has been removed, and any service connections have been reopened.

ularly well suited for repairing pipelines located under existing structures or large trees. It also is particularly useful for repairing pipelines located under busy streets or highways where traffic disruption must be minimized. Because the liner expands to fit the existing pipe geometry, this method is applicable to egg-shaped, ovoid, and arched pipelines. This method reportedly is effective for resolving corrosion problems. Inversion lining is also applicable in sewers needing minor structural reinforcement.

Inversion lining may be used for misaligned pipelines or in pipelines with bends where realignment or additional access is not required. Inversion lining also can be cost-effective where groundwater, soil conditions, numerous side connections, and surface structures or feature developments significantly increase the costs of other rehabilitation methods.

Inversion lining using water to cure the resins is generally confined to pipelines with diameters less than 1450 mm (57 in.) and manhole-to-manhole segments with distances less than 300 m (1000 ft) between manholes. Larger diameter pipelines, to an equivalent of 2750 mm (108 in.), have been lined by inversion techniques using air instead of water.

Limitations. As with other methods that focus on main line repair, inversion lining can merely transfer groundwater entry to defective service connections or other sewer sections not rehabilitated. Because this is a patented system available only through licensed contractors, relatively few contractors are available to use this system. Therefore, a prospective user may have difficulty securing competitive bids. In addition, standard specifications other than those provided by the manufacturer are not available to the prospective user.

Cost-effectiveness. Inversion lining is relatively new in this country. Thus it is difficult to determine how cost-competitive this method will be after the installed costs stabilize. This process has been used in Great Britain since 1970.[6]

MANHOLE REHABILITATION

Manholes are usually rehabilitated to correct structural deficiencies or to eliminate the entrance of surface water or groundwater. Manhole rehabilitation also may minimize or prevent corrosion of the internal surface caused by sulfuric acid. Sulfuric acid is formed when hydrogen sulfide gas is released from the wastewater into the humid sewer environment.

Many methods to rehabilitate manholes are available. Each method should be evaluated considering the type or types of problems and the physical characteristics of the structure, such as the condition, age, and type of original construction. For example, the methods used to rehabilitate brick manholes usually are different from those for precast concrete manholes.

EVALUATION/REHABILITATION

Likewise, sulfuric acid corrosion demands a treatment entirely different from preventing surface water from draining below the frame.

Frame and cover rehabilitation. A common problem with manhole covers and frames is the entry of surface water. Surface water enters through holes in the lid, through spaces around the lid between the frame and the cover, and under the frame if it is poorly sealed.

One solution to this problem is to install bolts through the holes. Either stainless steel bolts used with caulking compound or neoprene washers installed on the bottom of the cover can be used. A simple solution in some cases has been to insert corks into the pick holes. Another relatively simple solution is to use commercially available inserts, such as those shown in Figure 4.8, which are installed between the frame and cover. The manhole lid inserts prevent water, sand and grit from entering the manhole while gas is allowed to escape through vents.

Most foundries produce self-sealing frames and covers. These covers rest on flexible gaskets bearing on the frames. Care must be exercised when opening and closing these lids to prevent tearing the gasket. One of the disadvantages of using these self-sealing covers is the difficulty of inserting the gasket into the groove. There may be little allowable tolerance in the width of the groove with the result that many gaskets fall out. Also, the bottom edge of the cover can be weak and may chip.

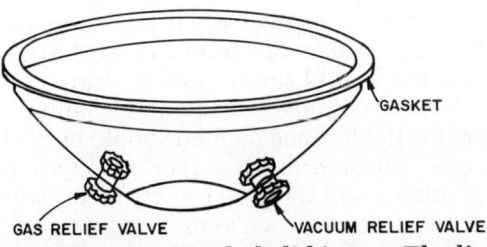

FIGURE 4.8. Manhole lid insert. The lip rests directly on the manhole frame. The manhole cover normally rests on top of the lip.

Deteriorated manhole frame and grade adjustment joints can be a significant source of inflow. Seals are quite often damaged because of road work, heavy traffic, freeze-thaw cycles, and snow plowing damage. Inflow is especially high if the manhole is subject to ponding. The need for repair materials and techniques depends on whether the frame must be raised. If a manhole frame is sound and properly graded but is poorly sealed to the manhole, one of several in-place rehabilitation methods can be used. None of these methods requires the expense and inconvenience of excavation.

Sealing frames in place may be performed by chiseling cracks and openings and applying hydraulic cement coated with a waterproofing epoxy. Oakum rope is sometimes used to fill large openings before applying the hydraulic cement. Freeze-thaw cycles may affect the lasting strength of the patch.

One of the best alternatives to minimize inflow through the cover, frame, or both is to raise the frame. To do this, manhole adjusting rings normally are used, although frame extension rings can also be used. The exposed exterior portion of the manhole also may be coated with cement mortar or a bituminous material.

Sidewall and base rehabilitation. Most rehabilitation work involving manhole sidewalls and bases is intended to minimize infiltration. Selection of a suitable rehabilitation method requires consideration of the physical condition of the manhole. If it is seriously deteriorated, replacing the manhole is often less costly. When the manhole is structurally sound, however, repair methods are normally used. Before the decision is made to rehabilitate or replace deteriorated manholes, the cause of the deterioration should be determined and the economies of alternative corrective measures identified.

Structural deterioration. Typically, severe deterioration is attributable to sulfuric acid corrosion, poor original construction, age, or overloading. Severe deterioration is usually resolved by replacing the manhole or by performing massive repair work. Corrective measures should either remove the causes of the deterioration or should include means to resist it.

In addition, rehabilitation should include measures to ensure manhole safety and efficient channel hydraulics. Manhole access ladder rungs are particularly important for safety. If the manhole is seriously deteriorated, then the manhole rungs are also suspect. For this reason, many communities and wastewater agencies do not install steps in new manholes. Weak rungs should be removed and not replaced, or a new corrosion-resistant rung should be installed.

The efficiency of the present channel should also be evaluated. If the flow is restricted or if disturbances are causing extraordinary head losses, repair work should improve the hydraulic characteristics. The existing base may have to be partly removed and reconstructed to provide better geometry, surface finish, or both. Existing flows are a problem when this is undertaken. Flows must be plugged temporarily and quick-setting products used, or flows must be temporarily routed around the structure being repaired. Flexible sleeves are also used to contain flows during repair.

Another important consideration is the entry requirement of maintenance equipment.[2,7] Cleaning tools, TV cameras and inline rehabilitation tools such as grouting packers, all require about 60 mm (24 in.) of straight pipe on grade for access. The channel should be built accordingly and self-cleaning benches should be provided.

Structural rehabilitation of manholes usually involves several steps and various products. Of primary importance is the removal of deteriorated materials, which can be accomplished by water or sand blasting or using mechanical tools. If a coating system is to be part of the rehabilitation, the entire surface should be thoroughly cleaned and prepared. The next step is to stabilize the remaining sound substrata using preparations designed for this purpose. These chemically stabilize the free lime and salts available from the cement and efflorescence. This should be done regardless of the manhole materials. Any surface irregularity such as missing bricks or spalled concrete should be patched. High strength moderately fast setting patching mortars greatly speed the work. To complete surface renewal, a lining or coating system may be applied.

A common denominator of all coatings is the necessity for a well-prepared surface that is free of active leakage. Some coatings require completely dry sidewalls.

The advantage of manhole repair work over replacement is that interference with traffic, other utilities, and sewer service is minimized. If the quality of the workmanship is high and the proper materials are selected, a satisfactory service life might be expected.

The economies of such rehabilitation depend on such factors as severity of chemical attack or corrosion, location, depth of manhole and water table, number of manholes requiring rehabilitation or replacement, and wastewater flow control measures needed. With precast manholes, however, it is usually possible to achieve the desired results at costs lower than that required for excavation and replacement.

In some situations, structural rehabilitation is not practical and replacement is needed. The details of manhole construction are widely known, and replacement should always include safety and operational considerations. Additionally, replacement often is preferable to other rehabilitation measures where permafrost or freeze-thaw cycles create special problems.

Point repairs. Often it is necessary to repair a limited length of pipe as it enters the manhole. This type of point repair requires care to avoid situations that would hinder the efficiency of the rehabilitation.

These repairs should utilize flexible compression-type couplings at the point where the pipeline connects to the manhole. The installation of this type of coupling often requires extensive excavation of the area near the manhole base. In this case, it is important to sufficiently compact the bedding material surrounding the pipe and manhole base. Failure to do so could result in differential settlement beyond the capacity of the flexible coupling or transfer of joint separation to points farther downstream.

Repairing sources of infiltration. Chemical grouting is the primary method used to eliminate infiltration, the most common manhole maintenance problem. Grouting can minimize infiltration through sidewalls and bases, around pipe entrances and drop structures, and under manhole frames. Chemical grouting is generally less costly than replacement or coating systems. Chemical grouting causes minimal traffic disruption, does not interfere with other utilities, is seldom affected by the wastewater flow, and requires no surface restoration work. Grouting should not be considered, however, where a structural solution is needed.[8,9,10]

Sidewall cracks normally are sealed by pressure injection grouting, which can be done in most cases with any of the basic types of grouts. Gel grouts have been used more in this application; however, urethane foam has been gaining wider acceptance. Regardless of the material used, the basic installation method is essentially the same, although the pumps and equipment differ.

Chemical grouting is generally the least costly method of eliminating infiltration into manholes, but the cost of rehabilitating individual manholes can vary widely. In some cases, infiltration rates have been minimized by chemical grouting at a low cost per manhole. In others, large quantities of grout are needed, resulting in high sealing costs.

Repairing superficial deterioration. Superficial deterioration of the manhole wall can eventually result in structural deterioration from exposure to the atmosphere. For example, sulfuric acid corrodes the cement used in concrete. In time, the steel in reinforced structures is exposed and the walls in nonreinforced structures crumble. Certain industrial wastes can also have a similar effect. If the corrosive atmosphere can be altered, then the process can be stopped before replacement is required.

A lining or coating system places a barrier between the concrete and corrosive atmosphere. The choice of material depends on the chemistry of the particular situation. Several materials, for example plastics and epoxys, are adaptable to this service.

Two other methods have been used—the inversion lining process discussed above and precut polyethylene or fiberglass sheets, which are fastened to the wall using caulked lap joints studded into the manhole wall.

SERVICE CONNECTION REHABILITATION

Service connections are pipelines that branch off the sewer main and connect building sewers to the public sewer main. Service connections may be as small as 100 mm (4 in.) in diameter, normally ranging from 4.5 to 30 m (15 to 100 ft) or more.

Service connections are built with any one of several products. Service connections are usually laid at a minimum self-cleansing grade from the building to the immediate vicinity of the main sewer. At this point the grade may change abruptly in order for the line to descend to the main sewer. Service connections normally enter sewer mains at angles ranging from 30 to 90 degrees from the axial flow direction and at vertical angles ranging from 0 to 90 degrees. In some developments, the same trench is used to route potable water service connections and the sewer service connection line. Consequently, any leaks in the potable water line can enter the sewer service connection line if it is not watertight.

The construction and maintenance of service connections is complicated by the fact that separate governmental agencies usually have jurisdiction over different portions of the sewers. The connection between the building's plumbing and drain system and the property line is often considered an extension of the in-structure facilities; therefore, it is ordinarily installed under plumbing or building codes and tested and approved by plumbing officials or building inspectors. The section of the building sewer between the property line and the street sewer, including the sewer main connection, usually is installed under sewer use rules, and inspection and approval are the responsibility of public works or sewer officials.

Industrial waste connections are often an exception to this rule. The construction of the entire length of many industrial service connections is usually supervised by collection system officials because of the possible effects of such wastes on sewer structures and treatment facilities.

For many years the effect of leaking service connections on the collection system and treatment facilities was considered insignificant. It was assumed that most service connections were above the water table and, therefore, were only subject to leakage during periods of excessive rainfall or exceedingly high groundwater levels. These sporadic conditions were not viewed as "serious" when compared to other collection system problems.

Today, the need to repair service connections is widely recognized. Recent I/I studies have demonstrated the impact of neglected building sewers on collection systems and treatment facilities. Research studies sponsored by EPA[11,12] indicate that a significant percent of the I/I in many collection systems is the result of defects in service connections. These defects include cracked, broken, or open-jointed pipes, which can allow storm-induced infiltration in addition to the longer-term infiltration not directly related to storms. Service connections also may transport water from inflow sources such as roof drains, cellar and foundation drains, basement or subcellar sump pumps, and "clean water" from commercial and industrial effluent lines. Service connections also can be a major source of infiltration when water migrates after sewer mains have been repaired.

To assess the potential for infiltration from service connections, the number of connections and the total length of the connection lines should be considered. Any given stretch of collection sewers in densely developed urban areas can contain multiple service connections. Furthermore, the total length of service connections is often equal to or greater than the sewer main length. For example, lots with a 15-m (50-ft) street frontage usually have four service connections per side of the street, or 8 per 60 m (200 ft) of block length. If the length of the average service connection is 8 m (25 ft) to the street line, the total length of these lines will be equal to the length of the sewer main.

In a survey of I/I control practices, representatives of state and provincial water pollution control agencies expressed opinions that illegal and poorly constructed service connections, improper sewer taps, and poor service connection construction practices were important sources of excessive I/I.

Although service connections probably contribute a large amount of the total infiltration carried by sewer systems, the exact extent has not been determined. To make such a determination, representatives from 26 agencies were interviewed during a national investigation.[11] The results from these interviews are listed in Table 4.1. Estimates of the percentage of infiltration in the total system attributable to service connections ranged so widely that the validity of any conclusions drawn from these data is questionable. The estimated percentages of the total infiltration attributable to service connections ranged from 95% in New Orleans and 75% in Baltimore to only 1% in Ft. Lauderdale, and a negligible amount in Nassau County, N.Y.

Estimates of the relative amount of infiltration from service connections were based on the assumptions that: a) the total length of service connections in the section of the sewer main used in the computation is twice that of the sewer main, and b) service connections have the same construction quality, in terms of tightness, as the sewer main system. These computations indicate that service connections could account for 38% of the total infil-

EVALUATION/REHABILITATION

TABLE 4.1. Estimated percentage of total infiltration attributed to building sewers.

City	Estimated Percentage
Baltimore, Maryland	75
Bloomington, Minnesota	25
Dallas, Texas	50
Denver, Colorado	High—no estimate
Ft. Lauderdale, Florida	1
Jacksonville, Florida	20
Knoxville, Tennessee	30
Milwaukee, Wisconsin	65
Nassau County, L.I., New York	Negligible
New Orleans, Louisiana	95
New Providence, New Jersey	0
Princeton, New Jersey	60
San Jose, California	60
Savannah, Georgia	35
Washington, D.C.	3
Washington Suburban Sanitary Commission	40
Watsonville, California	2
Yakima, Washington	40

Source: Reference 11.

tration into the entire sewer system. For Baltimore, where "row house" construction is common, 75% of the infiltration could result from service connections.

Rehabilitation methods

Chemical grouting. Three chemical grouting methods are currently available for sealing building sewers: pump full method; sewer sausage method (patented process); camera-packer method (patented process).

Pump full method—This method involves injecting a chemical grout through a conventional sealing packer from the sewer main up the service connection to an installed plug. As the grout is pumped under pressure, it is forced through the pipe faults into the surrounding soil where a seal is formed after the gel has set. After the sealing has been accomplished, excess grout is augered from the building sewer and the sewer is returned to service.

Sewer sausage method—This method is similar to the pump full method in that it requires access to the building sewer, the use of a camera-packer unit in the sewer main, and the injection of grout from the sewer main up the service connection to seal the pipeline. The primary difference is that a tube is inserted into the service connection before sealing to reduce the quantity of grout used and to minimize the amount of cleaning required after the sealing has been completed. The grout is pumped under pressure around the tube, up the service connection, and through any pipe faults into the surrounding soil where the seals are formed after the gel sets.

Camera-packer method—Unlike the other methods described, this method does not require placing equipment in the sewer main. It also differs in concept, as only faults seen through a television camera are repaired. First, a miniature television camera and a specialized sealing packer are inserted into the service connection. Using a tow line previously floated from the service connection access to the downstream manhole of the sewer main, the camera-packer unit is pulled into the service connection. The camera-packer is then slowly pulled back out, repairing faults that are seen through the television camera. Thus, the deepest leaking joints are sealed first. Joints and cracks are sealed in a manner similar to the conventional methods used for sealing joints in sewer mains. Once

REHABILITATION METHODS

TABLE 4.2. Miscellaneous private property rehabilitation measures.

Inflow Source	Possible Rehabilitation Measure
Connected downspout	Plug service connection opening and redirect downspout
Connected storm sump	Repipe to grade
Connected storm sump with diverter valve	Remove diverter valve
Defective or broken cleanout	Repair or replace as necessary
Connected area drain	Disconnect drain and install new sump pump
Connected crawlspace drain	Seal drain and install new sump pump
Connected foundation drain	Disconnect drain and install new sump pump

the repairs have been completed, the equipment is removed and the service connection returned to service.

The costs for this method vary, depending on the difficulties encountered when repairing service connections. When estimating costs, allowances should be made for such things as difficult site access and excavation dewatering.

Inversion lining. Limited experience with inversion lining of service connection lines indicates that the basic technology for sealing lines as small as 100 mm (4 in.) has been developed. As with sewer mains and laterals, inversion lining should reduce infiltration and, to some degree, improve the structural integrity of the existing pipeline.

The steps for lining a service connection using the inversion process are similar to those for lining a sewer main. An access point requiring excavation, however, is usually needed on the upstream side of the service connection line. Another variation from sewer main installations is the use of a special pressure chamber to provide the needed pressure to invert the fabric material through the service pipeline. The fabric also is terminated at the entrance of the sewer main, instead of at a downstream manhole.

After the curing process is completed, the downstream end of the liner is opened by excavation or via a remotely controlled cutting device placed in the sewer main. The upstream end is trimmed and the newly lined pipe is connected to the rest of the existing service connection line, restoring sewer service. The excavated soil is then replaced and all equipment is removed, completing the process.

Other rehabilitation measures. The methods discussed above to rehabilitate service connection lines should reduce infiltration resulting both from high groundwater levels and from rainstorms. In addition to these repairs, efforts should be made to remove sources of inflow that may be connected to the service connections. Inflow sources connected to service connection lines are usually located on private property. Therefore, a public awareness/public relations program often is needed. Such programs are intended to persuade property owners (without threat of legal consequences) to make the needed repairs to help correct a community problem.

Table 4.2 lists typical types of inflow sources found on private property and possible rehabilitation measures.

REFERENCES

1. "Gravity Sanitary Sewer Design and Construction." Manuals and Reports on Engineering Practice, No. 60, Am. Soc. of Civil Engineers, New York, N.Y. Manual of Practice No. FD-5, Water Poll. Control Fed., Washington, D.C., (1982).
2. "Operation and Maintenance of Wastewater Col-

lection Systems." Manual of Practice No. 7, Water Poll. Control Fed., Washington D.C., (1980).
3. "Sewer System Evaluation, Rehabilitation, and New Construction--A Manual of Practice." EPA-600/2-77-017d, U.S. EPA, Washington, D.C., (1977).
4. "Economic Analysis, Root Control and Backwater Flow Control as Related to Infiltration/Inflow." EPA-600/2-77-017a, U.S. EPA (1977).
5. Ouellette, H., and B.J. Schrock, "Rehabilitation of Sanitary Sewer Pipelines." Trans. Eng. J. ASCE, **107,** No. TE4, 497–513, (1981).
6. "Restoration of Sewerage Systems." Proceedings of an International Conference organized by the Institution of Civil Engineers, London, June 22–24, 1981. Thomas Telford LTD, London, (1982).
7. "Design & Construction of Sanitary and Storm Sewers." Manuals and Reports on Engineering Practice No. 37, Am. Soc. of Civ. Eng., N.Y.; Manual of Practice No. 9, Water Poll. Control Fed., Washington, D.C. (1969).
8. "Recommended Specifications for Sewer Collection System Rehabilitation." Nat. Assoc. of Sewer Service Companies.
9. "3M Sealing Gel System—Field Manual." (1980).
10. Conklin, G. F., and Lewis, P. W., "Evaluation of Infiltration/Inflow Program." EPA Project No. 68-01-49-13, U.S. EPA, Municipal Construction Division, Washington, D.C. (1980).
11. American Public Works Association, "Control of Infiltration and Inflow into Sewer Systems." 11022EFF12/70, U.S. EPA, Washington, D.C., NTIS PB 200 827 (1970).
12. Sullivan, R., et al., "Sewer System Rehabilitation and New Construction." EPA-600/2-77-017d, U.S. EPA, Washington, D.C., NTIS PB 279 248 (1978).

Chapter 5
MATERIALS USED FOR SEWER REHABILITATION

83   Manhole Rehabilitation Materials
     Manhole Covers
     Manhole Frame Grade
        Adjustment Joints
     Adjusting Rings
     Corbel and Wall Repair
        Materials
     Pipe Seal, Bench, and Channel
        Repair
87   Materials for Rehabilitation of
     Main Sewer Line
     Relining Materials
     Pipe Joint Materials
     Chemical Sealing Materials
92   References

New materials are continually being developed for use in rehabilitation of sewer systems and appurtenances. Sewer system rehabilitation presents conditions that may require materials different from those normally used for new sewer construction. This chapter provides information on various materials that are available for sewer rehabilitation.

Some of the factors to be considered in the selection of sewer rehabilitation materials are flow characteristics, durability, installation procedures, availability, and cost. Selection of a material should be based on specific requirements.

MANHOLE REHABILITATION MATERIALS

Manhole covers. Inflow entering the sanitary sewer systems through manhole cover pickholes and manhole frames and covers may be reduced by replacing existing manhole covers. Concealed pickholes do not extend completely through the manhole cover preventing inflow but still allow for removal of the cover with conventional manhole hooks.

A cover with concealed pickholes and a well-machined cover and frame joint seems to be the best alternative for a normal cover replacement condition. In some cases, it may be better to replace both the frame and cover at the same time. Because a large variety of frames and covers have been installed in the past, new covers may not be available to match the diameter, depth, and taper of an old frame that is excessively worn.

Manhole frame and cover joints may be sealed with appropriate gasket material. Self-sealing manhole covers have a tapered machine groove around the bottom edge into which a flexible gasket is inserted manually into the groove. The cover with the gasket inserted is then set on the frame lip with the weight of the cover compressing the gasket, thus making the seal. Figures 5.1 and 5.2 show manhole covers with concealed pickholes and self-sealing gaskets. Manhole insert "dishes" are discussed in Chapter 4.

Manhole frame grade adjustment joints

Sealing of raised frames. If a manhole frame must be raised, several materials may

EVALUATION/REHABILITATION

FIGURE 5.1. Manhole covers with concealed pickholes.

be used for providing a watertight seal. Waterproof additives, when added to backfill material, can prevent the flow of water into the joints.

A frame extension ring is designed to eliminate frame excavation for road resurfacing. Extensions are available that coincide with the desired height, fit existing frames, and are sized so that the existing cover will fit in the frame extension. Frame extension rings are made of cast iron, ductile iron, or steel, and increase the height of the frame in increments up to about 100 mm (4 in.). They are held in place by set screws or by expansion bolts and are used with watertight caulking material or sealing gaskets between the existing frame and the ring. Another type uses round or rectangular inner and outer rings secured by

FIGURE 5.2. Manhole covers with self-sealing tapered groove and gasket.

epoxy to the old frame. Manhole extension rings that can be adjusted after installation are also available. If the height or slant of the cover must be adjusted to match street elevations, the cover can be adjusted by turning adjusting bolts with a wrench handle provided with the cover. External sleeves may be utilized to extend the frame below pavement surfaces to avoid supporting the pavement weight. The basic concept is to anchor the bottom piece to the manhole chimney; meanwhile the other piece is free to move vertically while an elastic watertight seal fills the space between them. There is concern that asphalt pavement could not support two-piece frames.

Sealing frames and grade adjustments when the manhole must be raised also may be accomplished by using flexible rubberlike gasket material and precast concrete adjustment rings. The gasket material is placed in two concentric rings in all of the joints, and the weight of the ring compresses the gasket to seal the joint. Excellent bonding strength has been exhibited by this type of material. When using flexible gasket material, it is sometimes difficult to establish a precise final rim grade in paved areas because the material compresses significantly. There is also concern that the gasket could become "spongy" under traffic loads. Placing concrete that extends to the corbel can prevent movement. Care must be taken, however, that the gasket material not be placed spirally on the joint surfaces. Figures 5.3 and 5.4 illustrate double ring placement of the flexible gasket material.

In-place sealing of manhole frames. If a manhole frame is sound and properly graded but is poorly sealed to the manhole, one of several specialized rehabilitation materials can be used. Flexible rubber sleeves compressed against the side of the frame and adjusting rings with stainless steel expansion rings are available.

Elastomeric sealants have proved to be a good material for sealing manhole parts subject to movement from freeze and thaw conditions. When used in conjunction with

REHABILITATION MATERIALS

Adjusting rings. Adjusting rings are defined here as materials used to raise the grade of existing manhole rim elevations. This does not include manhole frame extensions, cones, or barrel sections. Adjustment rings are generally made of precast concrete, but they also are made with brick or concrete blocks. Adjustment rings are critical for achieving a level grade especially in traffic areas. To accommodate these needs, manufacturers have developed several types of adjustment techniques.

Flattops are reinforced concrete sections that may be set on a manhole barrel section; they eliminate the need for a corbel section because they have an eccentric hole to align with required adjustment rings and the frame. Flattops are useful for brick manhole rehabilitation work when corbel areas are partially defective.

An excellent height adjustment material frequently used for matching steep road grades is an inclined adjustment ring. Top and bottom surfaces are not parallel, which is ideal for providing firm support to a frame that must be pitched to match a road surface. An illustration of an inclined adjustment ring is shown in Figure 5.5.

Corbel and wall repair materials

Cement mixtures. These repairs are generally accomplished by applying high-strength compounds to the manhole walls and flow channels. These high-strength

FIGURE 5.3. Double ring placement of flexible rubber-like gasket material for adjustment of ring joints.

a bond breaker (for example, cellophane tape, foam rubber weather strip, or grease) applied over lines of suspected movement, elasticity of the material has maintained a watertight seal after exposure to winter conditions. Elastomeric sealants may be applied to both the interior and the exterior of the manhole. Care must be taken, however, to prevent puncturing of the sealant.

FIGURE 5.4. Double ring placement of flexible rubber-like gasket material for outside drop manhole joints.

FIGURE 5.5. Inclined adjustment rings.

EVALUATION/REHABILITATION

compounds are generally made by applying additives to cement mortar mixes that produce what is commonly referred to as hydraulic cement. Additives such as Portland cement, silica aggregates, plasticising agents, accelerating agents, and acrylic polymers are used to produce different types of hydraulic cement. Hydraulic cement is applied using trowels or brushes.

Other types of high strength compounds consist of hydraulic cement mixes with plaster additives, bonding agents, coal tar compounds, and many other specifically designed additives to meet different repair or patching requirements.

FIGURE 5.6. Flexible manhole sleeve and flange.

Manhole corbel and wall replacement. Generally, defective manhole corbels and walls will be replaced with precast concrete sections, which are readily available from many manufacturers. Tongue and groove joints provide for quick installation. Precast manhole sections are moved about with lifting holes previously cast in place or with outside lifting rings.

Reinforced fiberglass plastic manholes are lightweight and are effective for sealing out groundwater. They should have concrete bases that extend at least 150 mm (6 in.) beyond the barrel to avoid uplift problems. Precast rings or other suitable materials may be used to adjust frame grades with reinforced fiberglass plastic manholes. Manhole components are most often assembled by dealers although assembly can be performed in the field. Bonding surfaces must be cleaned with solvents, allowed to dry, and then joined with adhesive. Pipe connections are made in the same manner and then secured in position with stainless steel hardware.

Pipe seal, bench, and channel repair. Groundwater can travel along pipe sections in the bedding material. If defects are present in the pipe seal, bench, and channel sections of the manhole, significant quantities of water—which would be cost-effective to remove—may enter the sewer system.

Flexible manhole sleeves. To avoid infiltration problems caused by differential settling of manholes and pipe, flexible manhole sleeves are manufactured to provide a watertight union between the manhole and pipe. They also may allow some deflection at the joint. Some of the sleeves are cast in the manhole walls and others are held in place with an expanded stainless steel band. Flexible manhole sleeves should conform with ASTM C923. Sewer pipe is simply inserted in the sleeve and secured with stainless steel clamps. Installation time is relatively short when compared with other construction techniques. Flexible manhole sleeves are shown in Figure 5.6 and Figure 5.7.

FIGURE 5.7. Flexible manhole sleeves with flange cast in manhole wall.

Precast manhole bench and channel sections. If a manhole is severely deteriorated throughout, then complete replacement may be indicated. This would include the bench and channel section. An alternative for replacing the bench and channel section is to install a prefabricated precast manhole section with the bench and channel(s) retrofitted to match the existing piping network(s). Installation time may be reduced with correspondingly reduced labor costs using these prefabricated sections. A major disadvantage of using prefabricated manhole bottom sections is that field measurements must be made before fabrication to minimize error. Prefabricated plastic manholes currently are available.

Another valuable benefit of prefabricated manhole bottom sections is that outside drop connections also may be added during fabrication thus reducing field labor time.

Manhole steps are available that are designed for minimum deterioration. They may be installed in existing manholes where old steps have corroded and are no longer safe. Manhole steps are available in various materials.

MATERIALS FOR REHABILITATION OF MAIN SEWER LINE

Relining materials. When a sewer is to be relined, several options are available. Listed below are some of the materials available.

Sliplining materials. Polyethylene pipe and fiberglass pipe are the most common materials used for sliplining. Polyethylene pipe is a high-density, high molecular weight material available in various pressure ranges with higher pressures available only in the smaller diameters. Polyethylene pipe used for sliplining is currently available in larger diameters, structural strengths (wall thicknesses), and wall profiles to satisfy most situations.

The pipe and installations should conform, where appropriate, to the following specifications:

- "Polyethylene (PE) plastic pipe, Schedule 40 and 80 based on outside diameter" ASTM D 2447.
- "Polyethylene (PE) plastic pipe SDR-PR based on outside diameter" ASTM F 714.
- ASTM F-585, Standard Practice for Insertion of Polyethylene Pipe into existing sewers.
- ASTM D-2657 Heat Joining of Polyethylene Pipe and Fittings.
- Section XI, Sliplining of Sewers of Recommended Specifications for Sewer Collection System Rehabilitation, issued by the National Association of Sewer Service Companies.

Several factors that must be considered when determining the proper pipe diameter and wall thickness for polyethylene pipe include: external hydrostatic pressure, axial bending, and radial deflection. Hydrostatic pressure is a factor when the groundwater level is above the pipe, either intermittently or continuously. Equation 1 may be used to determine the standard dimension ratio (SDR) of pipe required:

$$SDR - 1 = F \sqrt[3]{\frac{E_a \left(\frac{2}{1-\mu^2}\right)}{P_h}} : F \sqrt[3]{\frac{2.508 E_a}{P_h}} \quad (1)$$

where:

- SDR = pipe standard dimension ratio, D_o/t where D_o equals outside diameter and t equals wall thickness in mm or in.,
- P_h = pressure due to head of water, kPa or psi,
- E_a = apparent (time-corrected) modulus (MPa × 10^3 or psi) for the grade of polyethylene used to manufacture the pipe,
- F = a design factor to account for "installed out of roundness," variability in the estimation of water head, etc. (F has a value less than 1.0), and equals C/N as defined below, and
- μ = Poisson's ratio (0.45 average value).

EVALUATION/REHABILITATION

The design factor F may be calculated from "out-of-roundness" or ovality values. C-values for ovality are:

Ovality, %	C
1	0.91
2	0.84
3	0.76
4	0.70
5	0.64

Ovality is generally taken as 1.5%. N is a design safety factor to compensate for variability in the estimation of water head and other uncertainties.

For pipe of non-homogenous wall structure such as pipe with external reinforcing ribs, the following form of Timoshenko's classic elastic buckling equation (from Equation 1) is derived and may be used to compute the minimum required pipe wall section stiffness for a given external hydrostatic loading:[1,2]

$$P_h = \frac{24 E_a I}{(1 - \mu^2) D_m^3} \cdot \frac{1}{F} \quad (2)$$

where

I = moments of inertia
D_m = mean diameter of pipe (in.)

and other symbols have been defined previously.

The choice of value (from manufacturer's literature) of E_a will depend on the estimated duration of the application of the P_h in relation to the design life of the structure. For example, if the total duration of the load P_h is estimated to be 25 yr, either continuously applied or the sum of intermittent periods of loading, the appropriately conservative choice of value for E_a will be that given for 25 yr of continuous loading at the maximum ground temperature expected over the life of the structure.

Fiberglass sliplining is limited to use in relatively straight sections of pipeline with diameters of 530 mm (21 in.) or more. The pipe longitudinally is not flexible and cannot be bent during installation procedures or curved around bends in the existing pipeline as polyethylene lining can be. This pipe can be cut with commercially available hand and power tools. Service connections are made by epoxy bonding saddles or attaching with stainless steel straps. The pipe is corrosion-resistant, comes in sections 6 m (20 ft) long, and has diameters of 510 mm (20 in.) or more.

Once the maximum anticipated hydrostatic pipe loading has been determined, it is a simple matter to specify a pipe product with the proper SDR. A series of preferred SDR numbers has evolved, with thinner wall pipe having a larger SDR number.

The amount of bending (temporarily during installation, or permanently) to which the liner pipe will be subjected is also important when determining size and weight. Polyethylene pipe is relatively flexible and may be curved during installation. Excessive bending, however, should be avoided. Longitudinal (axial) bends resulting from installation, or permanent bends to accommodate line or grade changes should be limited to radii equivalent to a longitudinal strain recommended by the pipe manufacturer. The minimum allowable radius of curvature for any size and weight of pipe can be approximated using the Equation 3:

$$R_c = \frac{D}{2 E_a} \quad (3)$$

where

R_c = radius of curvature, mm or in.,
D = diameter of the inserted pipe, mm or in., and
E_a = allowable axial strain.

Radial deflection at bends is automatically controlled if the radius of curvature is maintained as stated above.

An allowable axial strain of 1.5% which is equivalent to a minimum bending radius of D/[2 (0.015)] or 33D, is recommended by some manufacturers.

For temporary bends encountered during installation, a bending-radius-to-outside-pipe-diameter ratio of 25 (2% strain) or more is recommended. Where the bend is to be permanent, a minimum ratio of 35 is recommended. If the deflection is beyond the limits of the liner materials, additional excavation may be required.

Inversion lining. Inversion lining is accomplished by using needle felt, of polyester fiber, which serves as the "form" for the liner. The use of these products requires that the sewer be taken out of service during the rehabilitation period. One side of the felt is coated with the polyurethane membrane and the other is impregnated with the thermosetting resin. The felt variables include denier, density, type of material, method of manufacture (straight or cross lap), and length of fiber. The physical properties of the felt and chemicals must be determined for the specific project and in cooperation with prospective contractors.

The liner expands to fit the existing pipe geometry and therefore is applicable to egg-shaped, ovoids, and arch sewers.

Inversion lining has been utilized on lines from 100 to 2700 mm (4 to 108 in.) in diameter. It is normally applicable for distances between manholes of less than 600 m (2000 ft) or where groundwater, soil conditions, and existing structures make open excavation hazardous or extremely costly. Inversion lining with water is generally confined to pipelines with diameters less than 1450 mm (57 in.) and manhole-to-manhole segments with distances less than 300 m (1000 ft) between manholes. Normally air pressure is utilized for inversion techniques on larger diameter pipe. Compared with other methods, this process is highly technical. Other technical aspects include resin requirements, which vary with viscosity, felt liner, ambient temperatures and the filler in the felt content; the effects of ultraviolet light on the resin and catalyst; and safety precautions for personnel and property.

The cost of the raw materials, handling system, and curing procedures are high but may be offset with shorter construction time and the elimination of surface restoration.

Reinforced thermoset resin (RTR) and reinforced plastic mortar (RPM) pipe lining. RTR and RPM pipe may be used for the relining of deteriorating sewers. The pipe is available in diameters up to 3700 mm (144 in.). Proper resin selection can provide resistance to most chemicals and sewer gases over a wide temperature range.

The size is selected to allow clearance for threading inside the existing pipeline, considering the effects of offsets, lateral protrusions, and the like. Generally, 6 m (20 ft) lengths with gasketed joints are used to minimize pit size and facilitate installation by pushing into the existing pipeline. Inside joints are available to minimize the annular space between the liner and the existing line.

The pipe should conform, where appropriate, to the following standard specifications:
1. "Standard Specification for Reinforced Plastic Mortar Service Pipe," ASTM D 3262.
2. "Standard Specification for Reinforced Thermoset Resin Sewer Pipe," ASTM D 4184.

Grout and mortar lining. Large deteriorating sewers may be lined with mortar. The mortar that may contain cement and epoxy may be troweled, brushed, or rolled. The interior of the sewer may be lined with a cement mortar with added epoxy. The epoxy will increase the flowability, aid compaction, and strength. It hardens to a tough semi-rigid, impact resistant surface and bonds to steel and concrete surfaces under water. It is free of chlorides and may be used with reinforcing steel. There are several epoxy mortars available.

Additional information may be obtained from the manufacturers and from the following specifications:
1. "Chemical resistant resin mortars" ASTM C 395.
2. "Aggregates for masonry grouts" ASTM C 404.
3. "Grout for reinforced and non-reinforced masonry" ASTM C 476.

Shotcrete lining. Shotcrete is a term used to designate pneumatically applied cement plaster or concrete. A gun operated by compressed air is used to apply the cement mixture. The water is added to the dry materials as it passes through the nozzle of the gun. The quantity of water is

EVALUATION/REHABILITATION

controlled within certain limits by a valve at the nozzle. Low water ratios are required under ordinary conditions. The cement and aggregate are machine or hand mixed and are then passed through a sieve to remove lumps too large for the gun.

When properly made and applied, shotcrete is extremely strong, dense concrete, and resistant to weathering and chemical attack. Compared with hand placed mortar, shotcrete of equivalent aggregate-cement proportions usually is stronger because it permits placement with low water-to-cement ratios.

For relining existing structures, the shotcrete should be from 50 to 100 mm (2 to 4 in.) thick depending on conditions and may be steel reinforced. The cross-sectional area of reinforcement should be at least 0.4% of the area of the lining in each direction.

The following specifications should be considered:
1. "Specifications for concrete aggregates" ASTM C 33.
2. "Specifications for materials, proportioning, and application of shotcrete" ACI 506.
3. "Specifications for chemical admixtures for concrete" ASTM C 494.

Additional information may be obtained from other publications.[3]

Precast concrete segmental lining. A system of relining old sewers using precast concrete segments is being used in Europe and has been introduced into the U. S. The liners are usually cast in three parts, the invert and two sides that meet at a joint in the top. They may be cast for any size or shape, but are usually 1 m to 2 m (3.3 to 6.6 ft) in diameter.

Segments are transported on rails through the new liner and are assembled in place. The annular void between the old sewer and the new liner is often grouted.

Pipe joint materials. A wide variety of pipe joints are available for the different pipe materials used in sanitary sewer construction and are discussed in "Gravity Sanitary Sewer Design and Construction."[3] A good pipe joint must be watertight, root resistant, flexible, and durable.

Acrylamide gel. Acrylamide gel is typically a mixture of acrylamide and NN-methylene-bisacrylamide monomers, catalyzed with TEA (Triethanolamine 85%) and AP (ammonium persulfate). To inhibit root penetration, 200 mg/L of 2,6-dichlorobenzonitrile is added.

There are many catalysts and mixtures of catalysts that may be used. For normal use, however, the catalyst system is composed of TEA (85%), AP, and KFe (potassium ferricyanide). AP is a granular material and a very strong oxidizing agent. It is the initiator that triggers the reaction and is therefore the last material to be added. The induction period (gel time) begins with its addition. Generally, it is dissolved in water and added as a solution through a separate pump or by gravity. KFe is a reddish, granular material that is used to control the reaction.

The gel time (or "set time" or "induction time") is primarily affected by the concentrations of the catalysts and the solution temperature. Generally, the higher the catalyst concentrations and the temperature, the shorter the gel time. The concentration of the AP, however, normally should be less than 3.0% (by mass of entire mix) because at higher concentrations the mix may be too acidic to gel. Other factors that affect the gel time include monomer concentration, pH, metal ions, salts, particulate matter, hydrogen sulfide and chemical composition of mixing water. By altering the concentrations of the catalysts, the gel times can be controlled from 5 to 500 seconds. A gel time of approximately 20 seconds is commonly chosen in sewer grouting.

Before gelation, the grouting mix has a viscosity very close to that of water. This allows it to penetrate into small leaks and cracks in pipe walls and to mingle with outside soil particles. The acrylamide gel formed from the solution is a translucent, rubbery, and elastic material. Under moist conditions, the gel is resistant to attack by microorganisms, dilute acids, alkalies, salts, and gases normally found in the ground. When the gel is formed in a soil matrix, the permeability of the soil is reduced. The degree of reduction of the permeability depends upon the extent to which the voids are filled with the gel. If they are completely filled, the gel-soil mixture is virtually impermeable.

If allowed to dry, the gel will shrink because of dehydration. In a gel-soil mixture, dehydration may cause shrinkage cracks that would not be rehealed even if the moisture content of the mixture is restored later. Chemical additives such as ethylene glycol, however, may be added to the grout to prevent dehydration. Diatomaceous earth may be added to the grout to increase unconfined compressive strength. Pumping and mixing facilities may require modification if these additives are applied.

The acrylamide base sealing materials generally conform to the following basic properties:
1. A controllable reaction time of from 5 to 500 seconds.
2. Viscosity that remains constant throughout the induction period.
3. The ability to tolerate some dilution and react in moving water.
4. Production of a continuous irreversible, impermeable stiff gel in the final reaction.
5. A gel that is not rigid or brittle.
6. Base compounds that may be varied considerably by additives to increase the strength, adhesion, solution density, and viscosity.

Acrylate polymer grout. Acrylate chemical grout consists of a mixture of water-soluble acrylate monomers and crosslinker, which when catalyzed and initiated, forms a flexible gel that is used to seal sewer joints.[4] The grout has low viscosity, low permeability, and is resistant to attack from chemicals and bacteria found in sewer lines. The major differences between acrylate and acrylamide grout is that acrylate grout exhibits one/one hundredth the toxic exposure of acrylamide grout.

Acrylate polymer grout consists of solutions of acrylate monomers and a liquid crosslinker. In field application, the acrylate grout is diluted with water, catalyzed, and initiated. Gel time is affected by temperature and the amount of inhibitor added in the grout solution.

Acrylate gels are resistant to most chemicals found in sewer lines, and the grout is unaffected by 10% solutions of alcohols, ketones, hydrocarbons, acids, and bases, based on 16-week testing. Under severe drying (desert conditions), ethylene glycol may be added to the acrylate and acrylamide grouts to minimize loss of water.

Urethane foam. The grout is a liquid urethane prepolymer, which when mixed with water, foams and then cures to a tough, flexible, and cellular rubber. The first stage of the reaction is referred to as the "foam time," "induction time," or "cream time" and the second stage is called "cure time," "set time," or "gel time." Both the foam time and the cure time are temperature dependent. Generally, the higher the temperature, the shorter the reaction times. An accelerator, which is a water soluble amine, is usually added to the mixing water to reduce the foam and cured times. Low shrinkage of the foam grout is due to the expansive nature of the cure reaction forming a stable cellular structure. Cyclic wetting and drying conditions do not substantially affect the grout. The grout is resistant to most organic solvents, mild acids, and alkalies.

Urethane foam sealing materials generally conform to the following basic properties:
1. A controllable cure time from 15 minutes at 4° C (40° F) to 5 minutes at 38° C (100° F) when reacted with water.
2. When an accelerator is used, reduction in cure time.
3. During injection, foaming and expansion which causes steadily increasing viscosity.

Urethane foam grout is designed to cure with and absorb water into the reaction mass. The water:grout ratio for urethane foam is typically 1:1. When mixed with its catalyst and cured, the urethane foam will expand 10 to 12 times its original volume if not confined. When the reaction has taken place, the result is a tough, flexible, rubber-like elastomeric gel that acts like a gasket sealing the leaking joint or crack itself.

Foam grout takes longer to apply and to cure than gel grout, but is equally resistant to acids and alkalies and to most other solvents. Because this type of grout is in the joint rather than outside the pipe, it is affected less by dehydration than either of the gels.

Polyurethane gel. Urethane gel and acrylamide gel are different in that water is the catalyst for the urethane grout material. The gel is designed to cure with and absorb water into the reaction mass. The grout compound and water ratio can be varied over a wide range, thus allowing different gel strengths. The ratio of water to grout used for sewer pipeline sealing is approximately 8:1, although the exact ratio needed depends on the specific application and the gel strength desired. The gel is resistant to degradation from the action of biological agents, most chemicals, and solvents. Gel shrinkage is held to a minimum by using additives in the water component of the grout. The urethane-based gel should have the following basic properties:

1. A urethane polymer that reacts with water and forms a gel.
2. A mixture of urethane polymer and water that has a controllable gel time.
3. A urethane polymer that is pumpable.
4. A viscosity of the urethane polymer/water blend that is relatively constant from the time of mixing to the instant of gelation.
5. A flexible, tough gel that has at least 100% elongation to break.
6. Strength and resistance of the gel to shrinkage that can be increased by addition of a specific emulsion of elastomeric synthetic polymer in water. This emulsion should have the following properties:

Does not adversely affect the reaction of a urethane polymer with water;

Good mechanical stability that is not affected by normal pumping and mixing operations;

Good compatibility with the polyurethane polymer so that flexibility and toughness of the gel is retained;

Fillers may be added to increase strength of the gel and viscosity of the mixture provided that site conditions are appropriate for their use.

REFERENCES

1. Meldt, R., "The Strength of HDPE Pipes for the Renovation of Pipelines by Sliplining." 5th International Conference on Plastic Pipes, University of York, G.B. (Sept. 1982).
2. Sanbe, E., et al., "The Statics of Rigid Polethylene Drain Pipes." Kunstotte, **64** (April 1974).
3. "Gravity Sanitary Sewer Design and Construction." WPCF Manual of Practice FD-5; ASCE Manual of Engineering Practice No. 60; Water Pollut. Control Fed., Washington, D.C. (1982).
4. Clarke, W.J., "Performance Characteristics of Acrylate Polymer Grout." From "Proceedings of Conference on Grouting in Geotechnical Engineering, ASCE, New Orleans, Calif. (Feb. 1982).

Chapter 6

EFFECTIVENESS OF SEWER REHABILITATION

93 Measurement of Effectiveness of I/I Control
94 Studies of Rehabilitation Effectiveness for I/I Control
96 Expected Effectiveness of Sewer Rehabilitation
   Effectiveness of Individual Rehabilitation Methods for I/I Control and for Maintaining or Increasing Structural Integrity
   Overall Sewer System Effectiveness for I/I Control
99 Continuing Sewer Maintenance
100 References

Sewer rehabilitation has been used to maintain the structural integrity of the sewer system and to reduce infiltration and inflow. Since the passage of PL 92-500 in 1972, far more emphasis has been placed on sewer rehabilitation in an attempt to reduce the loads on wastewater treatment plants from excessive I/I.

This discussion of effectiveness includes information on both the ability of rehabilitation methods to reduce I/I, and the ability to improve or maintain structural integrity. Inflow control frequently does not actually involve sewer rehabilitation, but information on the effectiveness of a number of inflow control methods for flow reduction also is included.

Compared with the information available on methods of sewer rehabilitation, very little is available on the effectiveness of these methods for actually reducing I/I. There now is evidence, however, that the overall reductions expected from rehabilitation programs are not being realized. There is also increasing understanding of the reasons why these reductions are not being realized, which should result in a better understanding of the effectiveness of rehabilitation programs. Data and discussion are presented on the effectiveness of specific rehabilitation methods in relation to I/I control and structural integrity, and on the effectiveness for I/I control that might be expected on a total sewer system or a part of such a system.

MEASUREMENT OF EFFECTIVENESS OF I/I CONTROL

A measure of effectiveness of a rehabilitation program for I/I reduction is needed to make decisions regarding rehabilitation versus increasing treatment plant capacity. Because of the constantly changing flows in a sewer system and the difficulty of measuring or estimating accurately sanitary wastewater flows, the measurement of I/I under the best conditions may be considered inexact. At this time, there is an important need for accurate data on rehabilitation effectiveness to verify methods that might be developed for making effectiveness predictions. Little accurate data exist for this purpose.

A measurement of the effectiveness of a rehabilitation program obviously requires accurate measurement of flow in the sewer system before and after rehabilitation.

EVALUATION/REHABILITATION

Methods for carrying out these measurements are described in detail in Chapter 3. The simplest and most common way to calculate the effectiveness of rehabilitation is to subtract flows after rehabilitation from flows before rehabilitation and to divide by flow before rehabilitation minus the dry weather flows. For such a calculation to have any meaning, however, it is necessary that in addition to accurate flow measurements, groundwater conditions and rainfall durations and intensities are similar for measurements of flow made before and after rehabilitation, or that a dependable method is available for normalizing flows to a consistent set of conditions. The importance of these factors in obtaining meaningful results cannot be overemphasized. If monitoring were carried out for a long period of time before and after rehabilitation, conditions could be assumed to average out, but seldom, if ever, would the monitoring period be sufficiently long for this to be true. Although groundwater would only be expected to have a small effect on inflow, by the definition of infiltration it must have a very significant effect on that source of flow.

Good correlation exists between the logarithm of rainfall intensity and logarithm of peak inflow, verifying that groundwater is not an important variable affecting inflow.[1] This method also provides a way of extrapolating results before and after rehabilitation to a similar rainfall intensity or storm recurrence interval. The method is not intended for predicting inflow reduction and requires about one rainy season of data. Other researchers used this method at two locations.[2] Although the data show scatter, the results are sufficiently good for use in measuring inflow control effectiveness.

A method has been developed for measuring effectiveness of removal of both infiltration and inflow that is at least somewhat independent of groundwater conditions.[3,4] In this approach, rainfall-induced infiltration (IN) is defined as minimum flow after rainfall minus minimum flow before rainfall. Similarly, a rainfall-induced peak inflow (IW) is defined as peak flow during rainfall minus average flow before rainfall. It is assumed that rainfall induced flows are reduced by rehabilitation in proportion to total infiltration and inflow. Effectiveness of rehabilitation is then given for infiltration and inflow respectively as:

$$\frac{\text{IN before rehab} - \text{IN after rehab}}{\text{IN before rehab}}$$

and

$$\frac{\text{IW before rehab} - \text{IW after rehab}}{\text{IW before rehab}}$$

Generally speaking, methodology for making accurate determination of effectiveness is in an early stage of development and should benefit greatly from further investigation. With care in observing groundwater conditions and character of storms before and after rehabilitation, it is possible, however, to obtain a reasonable measurement of the effectiveness of a rehabilitation program.

Although flow monitoring is discussed in Chapter 3, it is important to point out here that monitoring for determining effectiveness must include a large enough part of the sewer system to ensure that sources eliminated in one location are not simply entering somewhere else by traveling through the bedding material surrounding the pipes to an area that has not been rehabilitated. For small systems, measurement of total flows at the treatment plant would assure meaningful results. For large systems, however, it is likely that some segment of the total system must be selected.

STUDIES OF REHABILITATION EFFECTIVENESS FOR I/I CONTROL

The most extensive survey ever conducted to determine specifically the effectiveness of sewer rehabilitation programs was carried out under support of the EPA.[5] The study was initiated as a result of evidence that I/I removal programs were not proving to be as effective as predicted and the programs were more time consuming and expensive than originally expected. Eighteen sewer systems were selected on the basis of the following: a) a Sewer System Evaluation Survey (SSES) was conducted, b) grouting was included in the rehabilitation program, and c) I/I reduction was reported to have been achieved. Observed reductions in I/I were much less than predicted by the SSES reports in all but one case. In three cases I/I increased. In the only community that experienced good I/I reduction, most of the infiltration resulted from leaky joints that were grouted successfully, and most of the sewer system was rehabilitated.

Some conclusions of the work follow:

- Excessive I/I was not generally eliminated;
- Post-rehabilitation I/I is exceeding treatment plant design for I/I components;
- The major sources of I/I in rehabilitated collection systems are building connections and unrehabilitated pipe joints;
- Major features in the I/I methodology are imprecise.

In addition to the EPA study there have been a number of other published studies that describe, in various ways, the effectiveness of sewer rehabilitation. There is also an increasing amount of information being collected in SSES reports that eventually may be published. Descriptions of effectiveness generally fall into one of three categories: a general comment on effectiveness unsubstantiated by flow or cost comparisons, studies that relate a cost savings or flow reduction but do not provide rehabilitation costs or other basis of determining cost-effectiveness, and those articles that make a cost comparison. Studies include both inflow and infiltration rehabilitation measures, and one article reports on infiltration rehabilitation of a storm sewer.

The body of information presently available[6-27] indicates there are effective methods for reducing the infiltration and inflow into sewers, but there are also problems and conditions that can exist in a sewer system and that can influence the overall effectiveness of I/I control. Service laterals, for which remedial measures are not well developed, represent a significant part of the problem. Not only do laterals contribute their own infiltration and often inflow, but frequently their connection to the main sewer causes leakage, which the usual grouting procedures cannot remedy. In addition, any joints near laterals cannot be grouted internally by the commonly used methods.

Another major source of poor effectiveness is the migration of water along pipes in the bedding material. When this can occur, any partial rehabilitation procedure is likely to be largely unsuccessful. Depending on grade, the water can move to unrehabilitated openings causing leakage to continue.[28] There is a good opportunity, in making a survey for I/I, that the I/I that would occur in the surcharged part of the system will be underestimated. The result will be rehabilitation of the unsurcharged part of the system. This will tend to reduce the flow in the system and cause some or all of the surcharged part of the system not to be surcharged any longer. That part of the system which would not have been adequately rehabilitated will then be susceptible to I/I and the observed effectiveness of rehabilitation will be less than predicted.

The migration problem, although not well defined, seems to be an important one that needs additional evaluation. Because dependable predictive means do not exist for this phenomenon and are not likely to be developed soon, careful consideration must be given to the possibility of migration in any rehabilitation program. Pilot rehabilitation studies of part of the system may be the only certain way to estimate the effect of migration.

The EPA study indicated that flows from house laterals were not adequately accounted for during performance of the infiltration inflow analysis.[5] By erroneously assuming that more of the I/I was from the main barrel than was actually true, too high an estimate of effectiveness for main barrel rehabilitation resulted.

EVALUATION/REHABILITATION

EXPECTED EFFECTIVENESS OF SEWER REHABILITATION

Effectiveness of individual rehabilitation methods for I/I control and for maintaining or increasing structural integrity. From the above discussions, it can be seen that there is only a moderate amount of information from which to judge confidently the effectiveness of rehabilitation methods for I/I control. There is additional information available from some manufacturers, but this is usually from tests conducted over a short time rather than over the life of a sewer line. In spite of the lack of certainty of results, engineers must continue to develop rehabilitation programs and estimate effectiveness as realistically as possible to make comparisons with costs of treatment.

The available information for major rehabilitation methods is summarized below. Included are comments on ability to improve structural stability. Descriptions of the methods are given in Chapters 4 and 5 and are not repeated here. Although inflow control frequently does not actually involve sewer rehabilitation, information on a number of inflow reducing methods is included.

Sewer replacement. The most expensive method of sewer rehabilitation is replacement. Where serious structural damage has occurred, this may be the only reasonable approach. Effectiveness for control of infiltration is obviously the same as for any other new sewer. Depending on the length of sewer replaced and the severity of migration of water outside the new pipe to the parts of the sewer adjacent to the replaced section, the realized infiltration removal will be less. Milwaukee, for example, uses 75% removal as a design figure for long sewers.[15] Reconnection of laterals will reduce the effectiveness further. Care must be used in making the reconnections to avoid leakage between the main barrel and the laterals, and an intense program of lateral rehabilitation must be carried out. Control of inflow will depend on how many inflow sources are eliminated during the new installation, but the opportunity exists for very effective exclusion of these sources. Obviously, such a procedure would have to be combined with a publicized inflow reduction program.

Sewer relining including inversion lining. Sewer relining involves placing a layer of piping material inside an existing pipe. Often this is done by inserting a slightly smaller pipe inside the existing pipe. Sewer relining may be cheaper than replacement and is more convenient because much less excavation is required. In the case of sliplining, however, the cost differential depends strongly on the density of laterals to be reconnected, and, it can be small if there is a large number of laterals.

There is a wide variety of lining materials ranging from cement applied directly to the inside of the existing pipe to modern plastic materials. (These are described elsewhere in this manual.) The information available indicates that slip-lining is an effective method for I/I control. The continuous plastic pipe linings have the potential for reducing infiltration to zero. Unless there are sufficient stresses to break the pipe, the control of infiltration should continue for the life of the pipe. Some of the other piping materials that are inserted but utilize the methods of joining pipe sections have a greater chance for leakage, but still should provide a high degree of infiltration resistance unless there is lateral displacement of the sections.

With the exception of thin coatings of mortar, the methods of sewer lining have the capability for maintaining or improving the structural stability of the sewer system and are often applied partly for that purpose. There is a range of materials available to meet structural needs when sliplining. The effectiveness for attaining structural integrity is increased by grouting completely the annular space between the new pipe and the original pipe. This procedure also ensures better infiltration control. The cost increase may, however, be significant. Steel reinforcing can be added when using mortar linings to increase their strength.

There are very few installations in the U.S. at this time for which reports of effectiveness have been published. Infiltration and exfiltration tests made in Northbrook, Illinois, on a 46 m (150 ft) and a 132 m (432 ft) segment of lined 300 mm (12 in.) vitrified clay pipe indicated zero leakage within the accuracy of the measurements.[30] If care is used in cutting out the part of the lining covering laterals, leakage around the laterals can be reduced to a very low value. Lack of care, on the other hand, could result in very poor infiltration control with much of the original leakage entering at these points. Because the method has been utilized only within the last decade, there is no long-term information on the useful life of this method. The Washington Suburban Sanitary Commission indicates that inversion lining can be accomplished at an average of 44% of the cost of replacement.[31] The cost relative to replacement is strongly affected, however, by the degree of difficulty and, therefore, the cost involved in excavating for pipe replacement. Inversion lining does not require excavation to reconnect laterals.

Inversion lining can be used for lining manholes and should exhibit the same high degree of infiltration reduction shown in sewer pipes. Care must be used in making openings to the sewers entering the manhole. Leakage at these points could reduce significantly the overall effect of lining.

The strength of the inversion liner can be varied by changing the thickness. As a result it can be made to overcome some structural instability. It has been used, for example, to stabilize brick sewers where the mortar was badly eroded and bricks had become loose or were even missing.

Sewer sealing. Many methods of sewer sealing have been used but little has been reported on the effectiveness of most of these methods. Since the 1960s there has been a sharp increase in the use of chemical grouts that can be applied as a freely flowing solution, but which solidify after application to form a flexible seal for leaking joints, cracks, and small holes.

An earlier section of this chapter cites a number of studies that presented information on the effectiveness of chemical grouting for infiltration control. It can be concluded from the available data that grouting is initially very effective in preventing infiltration, approaching 100% when done carefully. It must be pointed out, however, that these data result from joints and cracks that were properly sealed. It is possible for problems to occur during rehabilitation that result in poor sealing and require regrouting. It also must be stressed strongly that the high degree of effectiveness applies only to the joints sealed, not necessarily to the section of pipe involved in rehabilitation. Leakage from service laterals, joints close to service laterals, and from defects not correctable by the sealing procedure or in pipe sections adjacent to the section being sealed can significantly reduce overall infiltration removal effectiveness. Failure to recognize the importance of these leakage sources frequently has been a factor in the over-estimation of the degree of infiltration control. Unfortunately, there is presently no method other than a pilot sealing study to determine accurately how effective this method will be in a given location. Results cited earlier from the Washington Suburban Sanitary Commission and the Mamaroneck Sewer District give an indication of how significant leakage sources can be, but the range of results is too broad to select a figure that might be generally applied.[20-22]

Because of their newness, it is impossible to obtain direct evidence on the useful life of the modern grouting material. The longest life test cited in this chapter is 6 years. Structurally stable manholes can be sealed very effectively with the available chemical grouting materials, however.

Sewer sealing does not improve the structural stability of the sewer system and should not be used in sewers that are in poor physical condition. Unless the pipes are structurally stable, an effective sealing job probably could not be done and new leaks would develop rapidly.

Service lateral rehabilitation. Service laterals can constitute a serious source of both infiltration and inflow. Chapter 4 indicates that as much as 75% of the infiltration is service lateral related. Chapter 4 describes approaches toward rehabilitating service laterals. Methods other than excavation and replacement have not been widely used at this time and results of effectiveness have not been reported. Although some methods including replacement should provide a high degree of infiltration control within the pipe itself, there is no guarantee that they will control deliberate sources of inflow, and in some cases, infiltration that occurs under the building flow. Inflow control would require an effective disconnection and enforcement program.

In addition to I/I from the laterals, there is frequently infiltration resulting from leaky connection of the lateral to the main sewer and leakage at main sewer joints close to the lateral. The latter joints cannot be sealed with the usual packing equipment; therefore, all of these sources of infiltration represent a problem for which a well-documented low cost solution is not available.

For both the total lateral and the connection to the main sewer there is active development of remedial methodology that could produce dependable and more cost-effective solutions in the near future.

Inflow control. The following information originated from pilot evaluation of inflow control methods in Milwaukee, Wis., was part of a comprehensive SSES.[15] The objective of the work was to provide effectiveness factors that could be extrapolated over the entire collection system.

Manhole covers containing vent and pickholes can be significant sources of inflow when they are located in the path of surface runoff. Replacement with a waterproof, gasketed cover is estimated to be 90% effective in reducing inflow.

Manholes frequently leak between the frame and corbel, especially if there is heaving of the pavement from freezing. Use of elastomeric sealants poured or troweled onto the outside of the manhole or elastic sleeves is estimated to be 90% effective in reducing the leakage. Application of an adhesive sealant to the interior of the corbel and joint beneath the flange of the manhole frame is estimated to be only 75% effective because this method still allows water to enter the space between the frame and corbel, increasing the chance for seal failure from frost action.

Catch basins connected to sanitary sewers can contribute very large amounts of inflow to the system. Plugging of the connection to the sanitary system and reconnection to a storm drain is estimated to be 90% effective in reducing inflow. The effectiveness is estimated to be less than 100% to compensate for migration of some of the water to other parts of the sanitary sewer system.

Area drains and downspouts or roof drains are frequent sources of inflow. Disconnection of these from the sanitary system with a combination of reconnection to a storm sewer or discharge to the ground surface are estimated to be 90% effective in reducing inflow, with the remaining 10% finding its way into the sewer system by other routes. Where a significant fraction of these sources are discharged on the ground surface, rather than being connected to a storm sewer, the inflow reduction is likely to be somewhat less. An effective local enforcement program to prevent reconnection is required to maintain the indicated degree of inflow control.

Sump pump and foundation drain connections to sanitary sewers represent other significant sources of inflow. Disconnection of these sources and reconnecting either to storm sewers or discharge to the ground surface was observed to result in about 75% inflow reduction. To maintain that degree of control would require an effective enforcement program to prevent reconnection.

Overall sewer system effectiveness for I/I control. Because of a lack of understanding of the factors affecting I/I removal effectiveness, overestimation of the removal for total sewer systems or parts of systems has occurred. Reasons for possible overestimation are discussed in an earlier section of this chapter. Because of the significant involvement of groundwater, the problem seems to be with infiltration more so than inflow. Unfortunately, there are not generally applicable and dependable approaches available for making accurate estimates of effectiveness.

Because of the lack of methodology to predict accurately the overall effectiveness of proposed infiltration control systems, the most certain approach for obtaining dependable results is to conduct pilot studies of the effectiveness of the major rehabilitation methods. Results then can be extrapolated to the remainder of the system. This is being done with increasing frequency, especially in large cities. Results of several were discussed earlier in this chapter. Without pilot data, the engineer must consider as realistically as possible sources of leaks, such as service laterals, that have not always been adequately accounted for in the past. In two specific cases of pilot studies of joint grouting—one in the Washington, D.C., area and one in Westchester County, N.Y.—the range in infiltration removal was from 23 to 66%, with much of the remaining infiltration being accounted for by service laterals and manholes.[20-22] This wide difference should be a warning of the variability that can be encountered in the field, should make clear the need to consider carefully all the possible infiltration sources, and should make obvious the need to be conservative in estimating the effectiveness for infiltration control when actual data are not available on leakage rates from all significant sources.

An increasingly accepted concept in control of infiltration is the need for completeness in eliminating all sources in the sewers being rehabilitated. This approach is essential for preventing migration from interfering with good reduction. Although there is controversy over the influence that migration can have over long distances, there is little doubt that for short distances it can cause significant problems. The realization of the need for completeness in infiltration control began with sewer sealing, but now must be extended to service laterals, manholes, and any other significant leakage sources. As a further extension of this concept, where migration occurs over long distances, more will be gained by thoroughly rehabilitating selected long segments of the system, rather than many short segments. The latter alternative could produce much poorer results for the same cost.

The problem of estimating with reasonable accuracy the effectiveness of remedial measures for overall inflow control over a sewer system is generally less difficult than for infiltration. This chapter has included information on some common inflow sources that can be applied system-wide. The exclusion of many inflow sources from sanitary sewers would seem to be 100% effective, but in many cases there is an opportunity for some of the water to reenter the sewer. Correction for reentry must be considered when estimating the effectiveness of remedial measures.

CONTINUING SEWER MAINTENANCE

Many of the sewer systems in the U.S. are in a poor state of repair. Usually this happens because a higher priority is given to the more visible community problems for the funds that are available. The result is a minimum effort to try to avoid serious sewer collapses and to minimize flooding from sewer stoppages. Even when a major rehabilitation program is undertaken, such as that required when building a new wastewater treatment plant, there is often little thought given to continuing sewer maintenance. Whether a system is new or has recently been rehabilitated does not mean that continuing maintenance is not necessary to ensure structural integrity and to prevent gradual increases in I/I. Many problems could undoubtedly be taken care of more cheaply if they were not allowed to worsen. With the generally recognized decline in federal support for sewer re-

EVALUATION/REHABILITATION

habilitation and reconstruction, it is more important than in the past that municipalities and other sewer agencies give greater consideration to improving preventive maintenance programs. Only by instituting such a program can these agencies prevent the needless, burdensome expense to the public that will eventually develop when the sewer system has reached the point where major rebuilding is required.

REFERENCES

1. Nogaj, R.J., and Hollenbeck, A.J., "One Technique for Estimating Inflows with Surcharge Conditions." J. Water Pollut. Control Fed., **53**, 4, 491 (Apr. 1981).
2. Nelson, R.E., and Bodner, R.L., "Measuring Effectiveness of Infiltration/Inflow Removal." Paper presented at American Public Works Congress, Atlanta, Ga. (Sept. 16, 1981).
3. Chatterjee, S., "Infiltration/Inflow Reduction in Chicago-Suburban Sewer System." Project Report, Citizens Utilities Company, Addison, Ill. (Sept. 1979).
4. Chatterjee, S., Private communications (July 1982).
5. Conklin, G.F., and Lewis, P.W., "Evaluation of Infiltration/Inflow Program." Project No. 68-01-4913, U.S. EPA, Washington, D.C. (1980).
6. Calhoun, T.P., "Longevity of Sewer Grout Under Severe Conditions." Public Works, **106**, 10, 80 (1975).
7. Kemmet, R.H., "Sewer Grouting Cures Plant Overload Problem." Public Works, **104**, 7, 94 (1973).
8. Sweeney, C.G., "Grout Routs Sewer Problems." Water Wastes Eng. **14**, 5, 59 (1977).
9. "A $100,000 Savings Through Pipe Relining." Plant Eng., **29**, 30 (February 6, 1975).
10. "Grouting Saves Sewer Line." Water Sew. Works, **122**, 11, 54 (1975).
11. "Grouting Provides Economical and Effective Maintenance in Kansas." Water Sew. Works, **125**, 3, 66 (1978).
12. Gutierrez, A.F., Private communications (March 1983).
13. Gutierrez, A.F., and Wilmut, C., "The Feasibility of Infiltration/Inflow Elimination." Public Works, **106**, 4, 68 (1975).
14. "Manhole Inserts Abate Inflow." Public Works, **110**, 12, 47 (1979).
15. "Milwaukee Metropolitan Sewage District Sewer System Evaluation Survey." Milwaukee MSD, Milwaukee, Wis. (Aug. 1981).
16. "Grouting Stops Infiltration in Large-Diameter Storm Drain." Public Works, **106**, 9, 82 (1975).
17. "Sealing Sewers in Unstable Soils." Am. City and County, **96**, 3, 48 (1981).
18. Wass, V.C., and Bush, C.M., "Benefits from Grouting an Entire Sewer System." Public Works, **108**, 11, 55 (1977).
19. Morgan, H.E., "How Long Does Sewer Joint Grouting Last?" Public Works, **105**, 10, 98 (1974).
20. Bonk, M.P., et al., "Test Program for the Anacostia Sewerage System Rehabilitation." Report No. 1, Washington Suburban Sanitary Commission, Hyattsville, Md. (October 1978).
21. Bonk, M.P., et al., "Test Program for the Anacostia Sewerage System Rehabilitation." Report No. 2, Washington Suburban Sanitary Commission, Hyattsville, Md. (May 1979).
22. "Sewer System Evaluation Survey Mamaroneck Sewer District." Department of Environmental Facilities, Westchester County, N.Y. (December 1982).
23. Rhodes, D.E., "Rehabilitation of Sanitary Sewer Lines." J. Water Pollut. Control Fed., **38**, 215 (1966).
24. Montgomery, Ohio, County Sanitary Department, "Ground Water Infiltration and Internal Sealing of Sanitary Sewers." 11020 DHQ 06/72, U.S. EPA, Washington, D.C. (1972).
25. Greenleaf, D.J., "Grouting Job Pays for Itself in Eight Months." Public Works, **109**, 12, 38 (1978).
26. "Sewer Rehabilitation Cuts Infiltration." Water Sew. Works, **124**, 11, 86 (1977).
27. Cesareo, D.J., and Field, R., "How to Analyze Infiltration/Inflow." Water Sew. Works, **122**, R, R84 (1975).
28. Steketee, C.H., "Why EPA's I/I Control Program Fails." Public Works, **112**, No. 7, 51 (1981).
29. Sullivan, R.H., et al., "Economic Analysis, Root Control, and Backwater Flow Control as Related to Infiltration/Inflow Control." EPA-600/2-77-017a, U.S. EPA, Washington, D.C. (1977).
30. Driver, F.T., and Olson, M.K., "Demonstration of Sewer Relining by the Insituform Process, Northbrook, Illinois." Grant No. R806322, U.S. EPA, Cincinnati, Ohio (1982).
31. Thomasson, R.O., "Repair Replace Reconstruct All in One." Washington Suburban Sanitary Commission, Hyattsville, Md.

Index

A
Acrylamide gel, pipe joint materials, 90,91
Acrylate polymer grout, pipe joint materials, 91
Adjusting rings, manhole rehabilitation, 85
Anemometer, velocity, 51
Animals, manhole hazards, 16
Annual volume, flow component, 8
Atmospheric hazards, manhole hazards, 16
Automatic meters, flow measurement, 54,55

B
Base, manhole rehabilitation, 76–78
Base flow, flow component, 7,8
Bench repair, manhole rehabilitation, 86,87
BOD evaluation, infiltration quantification, 28
Bubbler, depth recorder, 54
Bucket test, flow measurement, 53
Bypass information, system records, 44
Bypasses,
   flow measurement, 32
   site selection, 58

C
C-values, ovality, 88
Calibrated discharge curves, flow measurement, 51
Camera-packer method, chemical grouting, 80
Capacitance/electronic, depth recorder, 55
Cement mixtures, manhole rehabilitation, 85,86
Channel repair, manhole rehabilitation, 86,87
Chemical grouting,
   manhole rehabilitation, 78
   pipeline rehabilitation, 67–71
   service connection rehabilitation, 80
Chemical tracers, flow measurement, 49,50
Chemical treatment, root control, 65
Cleaning, pipeline, 24–27
Coatings, pipeline rehabilitation, 71
Collection lines, rainfall-induced infiltration/inflow, 35
Concrete sections, manhole rehabilitation, 86
Contained flows, infiltration/inflow quantification, 32
Corbel, manhole rehabilitation, 85,86
Cost,
   chemical grouting, 70,71
   inflow elimination, 37
Cost-effectiveness,
   infiltration location, 31
   infiltration/inflow localization, 36,37
   inversion lining, 75
Covers,
   manhole rehabilitation, 76
   rehabilitation materials, 83,84
Current meter, velocity, 51
Curvature radius, equation, 88

D
Daily flow comparison, infiltration quantification, 29
Damage identification, dye-water flooding, 18
Data, flow measurement, 33–35
Data evaluation,
   evaluation procedures, 9
   groundwater gauging, 11,12
Data needs, flow monitoring, 43–45
Data recording,
   inspection, 17
   television inspection, 25
Data storage, flow metering, 8
Defect indicators, O&M problems, 39
Demography, system records, 44
Depth, automatic flow meters, 54

Differential isolation, isolation technique, 22,23
Doppler meters, velocity, 55
Drowning, manhole hazards, 16
Dye dilution, flow measurement, 24,49
Dye interval timing, flow measurement, 23,24
Dye-water flooding,
  evaluation technique, 17–20
  leak quantification, 26

E

Elastic buckling, equation, 88
Elastomeric sealants, manhole rehabilitation, 84,85
Electromagnetic meters, velocity, 55
Emergency pumping, system records, 44
Environmental Protection Agency, 1,2,6,79,95
Equipment,
  dye-water flooding, 17
  groundwater gauging, 11
  pipeline cleaning, 24,25
  precipitation measurement, 10
  site selection, 59
  smoke testing, 12
  television inspection, 25
Escaping flows, infiltration/inflow quantification, 32
Evaluation techniques, program design, 21
Excavation, pipeline rehabilitation, 66
Excessive infiltration/inflow, definition, 3
Expression of results, flow isolation, 24

F

Fiberglass,
  manhole rehabilitation, 86
  sliplining materials, 88
Flexible sleeves, manhole rehabilitation, 86
Float, depth recorder, 54
Flow components, sewer evaluation, 7,8
Flow data, evaluation procedures, 9,10
Flow measurement,
  infiltration/inflow quantification, 32,33
  rehabilitation effectiveness, 93,94
  velocity, 51

Flow metering, evaluation procedures, 8,9
Flow monitoring, program design, 56–62
Flow parameters, flow measurement, 33
Flow records, collection system, 43
Flume, H, 49
Flume, Palmer-Bowlus, 48,49
Flume, Parshall, 48
Flume, trapezoidal, 49
Flumes, flow measurement, 46–49
Fluorometric method, flow measurement, 24
Frame, manhole rehabilitation, 76
Frame extension ring, manhole rehabilitation, 84
Frame grade adjustment joints, manhole, 83–85
Freshwater data, system records, 44

G

General considerations, sewer rehabilitation, 63,64
Groundwater gauging, evaluation techniques, 11
Groundwater levels, infiltration location, 31
Groundwater measurement, flow measurement, 55,56
Groundwater migration, night flow isolation, 21
Grout, sliplining material, 89
Gunite, pipeline rehabilitation, 71

H

Hazardous chemicals, manhole hazards, 16
Hazards, inspection, 16,17
Heat fusion saddles, sliplining, 74
Herbicides, root control, 65
Hydrographs, data evaluation, 10

I

In-place sealing, manhole frames, 84,85
Incremental infiltration, definition, 30
Indirect inflow, leak quantification, 26
Infection, manhole hazards, 16
Infiltration,
  definition, 2,7
  evaluation, 27–31

flow component, 7,8
leak quantification, 26
service connection, 79,80
Infiltration sources, manhole rehabilitation, 78
Infiltration/Inflow,
  definition, 3
  detrimental effects, 6
  hydraulic load, 1
Infiltration/Inflow analysis, sewer evaluation, 5,6
Infiltration/Inflow control, expected effectiveness, 98,99
Infiltration/Inflow control effectiveness, measurement, 93,94
Infiltration/Inflow determination, dye-water flooding, 18
Infiltration/Inflow verification, dye-water flooding, 19
Inflow,
  definition, 2,7
  evaluation, 32
  flow component, 7,8
  infiltration quantification, 27
Inflow control, expected effectiveness, 98
Inflow identification, dye-water flooding, 18
Insects, manhole hazards, 16
Inspection, procedures, 15
Interpretation of results, flow isolation, 24
Interviews, infiltration location, 31
Inversion lining,
  expected effectiveness, 96,97
  pipeline rehabilitation, 74,75
  service connection rehabilitation, 81
  sliplining material, 89
Isolation techniques, program design, 21,22

J

Jurisdiction, service connection, 78,79

L

Leak quantification, television inspection, 25–27
Legislation, PL 92–500, 1,2,6
Lift stations, flow metering, 9
Linings, pipeline rehabilitation, 71

Localization, rainfall-induced infiltration/inflow, 35,36
Location, infiltration evaluation, 30,31
Location techniques, inspection, 14,15

M

Maintenance, monitors, 60
Manhole inspection, physical evaluation, 40
Manholes,
  leak quantification, 26
  rainfall-induced infiltration/inflow, 35
  site selection, 57
Manning's equation, open channel flow, 26,50
Manual methods, flow measurement, 45–54
Map accuracy, physical conditions, 40,41
Maps,
  data needs, 43
  location technique, 14
  sewer records, 43
Materials,
  manhole rehabilitation, 83,84
  sewer rehabilitation, 83–92
Measurement accuracy, night flow isolation, 20,21
Measurement techniques,
  flow isolation, 23
  flow monitoring, 45–56
Meter, doppler ultrasonic, 51
Meter maintenance, evaluation procedures, 9
Methodology, infiltration/inflow localization, 36
Methods, service connection rehabilitation, 80,81
Monitoring,
  continuous long-term, 62
  continuous short-term, 61,62
  duration, 61,62
  flow, 43–62
  instantaneous, 61
  permanent, 62
  program design, 60
  random interval, 61
Monitoring site, hydraulic characteristics, 9
Mortar lining, sliplining material, 89
Multi-pass, isolation method, 21

103

N

Night flow isolation, evaluation technique, 20,21
Nighttime domestic flow, infiltration quantification, 29
NOAA, data source, 10,11
Nozzle meters, velocity, 55

O

O&M problems, physical conditions, 39
Orifice meters, velocity, 55
Ovality values, pipe, 88
Overflow information, system records, 44
Overflows,
    flow measurement, 32
    site selection, 58

P

Peak rate, flow component, 7
Photographs, television inspection, 25
Physical conditions, evaluation, 38–41
Physical injury, manhole hazards, 16
Piezometer, gauging equipment, 11
Pipe, structural integrity, 38
Pipe joint materials, sewer rehabilitation, 90–92
Pipe seal repair, manhole rehabilitation, 86,87
Pipeline inspection, physical evaluation, 40
Plug inflation, plugging, 22
Plug removal, plugging, 22
Plugging, isolation technique, 21,22
Point repairs,
    manhole rehabilitation, 77,78
    pipeline rehabilitation, 66,67
Polyethylene, sliplining, 71
Polyurethane gel, pipe joint materials, 92
Pre-installation considerations, evaluation procedures, 8
Precast concrete, sliplining material, 90
Precast sections, manhole rehabilitation, 86,87
Precipitation measurement, 10,11
Pressure sensor, depth recorder, 54
Primary control devices, flow metering, 9,10

Probe,
    depth recorder, 54
    doppler ultrasonic, 51
    magnetic, 51
    velocity, 51
Procedures,
    dye-water flooding, 18
    evaluation techniques, 8
    smoke testing, 14
    television inspection, 25
Public notification, smoke testing, 13
Pump full method, chemical grouting, 80
Pump station, flow measurement, 52
Pumps, calibration, 52,53

Q

Quality control, monitoring data, 60
Quantification, rainfall-induced infiltration/inflow, 32–35
Quantification techniques, infiltration evaluation, 27–30

R

Rainfall data, system records, 44
Rainfall measurement, flow measurement, 56
Rainfall-induced infiltration, definition, 7,94
Rainfall-induced infiltration/inflow, evaluation, 32
Rainfall-induced peak flow, definition, 94
Raised frames, manhole rehabilitation, 83,84
Reach length, night flow isolation, 20,21
Record keeping program, system records, 44
Rehabilitation,
    manhole, 75–78
    pipeline, 64–75
    service connection, 78–81
Rehabilitation effectiveness,
    expected, 96–99
    infiltration/inflow control, 96–98
    structural integrity, 96–98
Reinforced plastic mortar, sliplining material, 89
Reinforced shotcrete, pipeline rehabilitation, 71

Reinforced thermoset resin, sliplining material, 89
Relining materials, sewer rehabilitation, 87–90
Remote connector fittings, sliplining, 73,74
Replacement, pipeline rehabilitation, 66
Root control, pipeline rehabilitation, 64–66
Root intrusion, pipeline rehabilitation, 65
Root removal, pipeline rehabilitation, 64–66

S

Safety precautions,
   dye-water flooding, 19
   inspection procedures, 15–17
   manhole hazards, 16,17
Safety program, program design, 59
Service connections,
   pipeline rehabilitation, 66
   sliplining, 72,73
Service lateral rehabilitation, expected effectiveness, 98
Service lines, rainfall-induced infiltration/inflow, 35
Sewer condition, system records, 43
Sewer evaluation, methods, 5–41
Sewer maintenance, rehabilitation effectiveness, 99,100
Sewer rehabilitation,
   effectiveness, 93–100
   methods, 63–81
Sewer relining, expected effectiveness, 96,97
Sewer replacement, expected effectiveness, 96
Sewer sausage method, chemical grouting, 80
Sewer sealing, expected effectiveness, 97
Shotcrete lining, sliplining material, 89
Sidewall, manhole rehabilitation, 76–78
Single-pass, isolation method, 21
Site selection, program design, 57
Siting,
   groundwater gauging, 11,12
   permanent, 59
   temporary, 59
Sliplining, pipeline rehabilitation, 71–74

Sliplining materials, sewer rehabilitation, 87,88
Small sub-systems, infiltration quantification, 29
Smoke testing, evaluation techniques, 12,13
SSES,
   effectiveness studies, 95
   sewer evaluation, 5,6
Stage discharge curve, monitoring data, 60
Stage measurement, flow measurement, 54
Standard dimension ratio, equation, 87
Structural deterioration, manhole rehabilitation, 76,77
Structural integrity, physical conditions, 38,39
Sulfuric acid corrosion, manhole rehabilitation, 75,76
Superficial deterioration, manhole rehabilitation, 78
Surcharging, infiltration/inflow quantification, 32
System records, data needs, 43

T

Tapping saddles, sliplining, 74
Techniques,
   infiltration/inflow localization, 36
   physical evaluation, 40
   sewer evaluation, 7
Television inspection,
   manhole, 41
   pipeline, 24–27,41
Temperature, infiltration location, 30,31

U

Ultrasonic, depth recorder, 55
Urethane foam, pipe joint materials, 91

V

Velocity,
   automatic flow meters, 55
   flow measurement, 51
Velocity-area method, flow measurement, 23,49,50
Venturi meters, velocity, 55

Venturi principle, flumes, 46
Videotape, television inspection, 25
Visual inspection,
  infiltration location, 31
  manhole, 14,40
  pipeline, 14,40

W

Wall repair, manhole rehabilitation, 85,86
Water usage, system records, 44
Water use evaluation, infiltration quantification, 27,28

Weir,
  compound, 46
  flow measurement, 23
  flow monitoring, 46
  leak quantification, 26
  rectangular (contracted) with end contraction, 45
  rectangular (suppressed) without end contractions, 46
  trapezoidal (Cipolletti), 46
  triangular (V-Notch), 45
Weirs, flow measurement, 45,46
Wet weather infiltration, definition, 32